T0174791

SEA MANAGEMENT.
A THEORETICAL APPROACH

To commemorate

the Quincentenary of the Discovery of the Americas

the Twentieth Anniversary of the United Nations
Conference on the Human Environment

the Tenth Anniversary of the United Nations
Convention on the Law of the Sea

Published with the collaboration of
ENTE COLOMBO '92, Genoa, Italy

Colombo '92
ESPOSIZIONE DI GENOVA

SEA MANAGEMENT.
A THEORETICAL APPROACH

ADALBERTO VALLEGA

Institute of Geographical Sciences
University of Genoa,
Genoa, Italy

CRC Press
Taylor & Francis Group
Boca Raton London New York

CRC Press is an imprint of the
Taylor & Francis Group, an **informa** business

A TAYLOR & FRANCIS BOOK

CRC Press
Taylor & Francis Group
6000 Broken Sound Parkway NW, Suite 300
Boca Raton, FL 33487-2742

First issued in paperback 2019

© 2005 by Taylor & Francis Group, LLC
CRC Press is an imprint of Taylor & Francis Group, an Informa business

No claim to original U.S. Government works

ISBN-13: 978-1-85166-772-7 (hbk)
ISBN-13: 978-0-367-86377-7 (pbk)

This book contains information obtained from authentic and highly regarded sources. Reasonable efforts have been made to publish reliable data and information, but the author and publisher cannot assume responsibility for the validity of all materials or the consequences of their use. The authors and publishers have attempted to trace the copyright holders of all material reproduced in this publication and apologize to copyright holders if permission to publish in this form has not been obtained. If any copyright material has not been acknowledged please write and let us know so we may rectify in any future reprint.

Except as permitted under U.S. Copyright Law, no part of this book may be reprinted, reproduced, transmitted, or utilized in any form by any electronic, mechanical, or other means, now known or hereafter invented, including photocopying, microfilming, and recording, or in any information storage or retrieval system, without written permission from the publishers.

For permission to photocopy or use material electronically from this work, please access www.copyright.com (http://www.copyright.com/) or contact the Copyright Clearance Center, Inc. (CCC), 222 Rosewood Drive, Danvers, MA 01923, 978-750-8400. CCC is a not-for-profit organization that provides licenses and registration for a variety of users. For organizations that have been granted a photocopy license by the CCC, a separate system of payment has been arranged.

Trademark Notice: Product or corporate names may be trademarks or registered trademarks, and are used only for identification and explanation without intent to infringe.

English linguistic assistance
Arthur Lomas
Hance D. Smith

Maps and graphics
Paolo Cornaglia
Nicoletta Varani

British Library Cataloguing in Publication Data

Vallega, Adalberto
Sea Management: A Theoretical Approach
I. Title
333.91

Library of Congress CIP data applied for

No responsibility is assumed by the Publisher for any injury and/or damage to persons or property as a matter of products liability, negligence or otherwise, or from any use or operation of any methods, products, instructions or ideas contained in the material herein.

Special regulations for readers in the USA

This publication has been registered with the Copyright Clearance Center Inc. (CCC), Salem, Massachusetts. Information can be obtained from the CCC about conditions under which photocopies of parts of this publication may be made in the USA. All other copyright questions, including photocopying outside the USA, should be referred to the publisher.

Visit the Taylor & Francis Web site at
http://www.taylorandfrancis.com

and the CRC Press Web site at
http://www.crcpress.com

PREFACE

This book has been conceived with the aim of contributing to the International Conference on Ocean Management in Global Change (Genoa, June 22-26, 1992) and to the ocean sciences' debate on the conceptual framework and targets of sea management. The former objective is justified by the background of the Conference, which is aimed at encouraging a multidisciplinary approach to the management of the sea and implementing comprehensive approaches to sea resource use and environmental protection and conservation. The latter objective is justified by the growing importance which literature has been attributing to the theoretical and methodological bases of sea management since the late Seventies.

The thesis which supports this work is that the more marine scientists are capable of moving towards a general system-based approach to sea uses and environmental implications, making holistic views, and creating common forms for implementing multidisciplinary views, the more sea management will have the opportunity of advancing.

My thanks are due to Stella Maris Vallejo and Moritaka Hayashi, United Nations Office for Ocean Affairs and the Law of the Sea, and Hance D. Smith, Department of Maritime Studies of the University of Wales (Cardiff), for their stimulating conversations about the background and goals of sea management, to Alberto Bemporad, the General Commissioner of the Specialized International Exhibition "Christopher Columbus: Ships and the Sea", and Giorgio Doria, the Co-ordinator of the Technical-Scientific Committee of the Ente Colombo 92, who encouraged me to offer this contribution to the scientific debate which is expected in the context of the Celebrations of the Quincentenary of the Discovery of the Americas.

Adalberto Vallega
The University of Genoa, Italy

ABBREVIATIONS

CAM	Coastal Area Management
CZM	Coastal Zone Management
DIESA	United States Department of International Economic and Social Affairs, United Nations
dwt	deadweight tons
EEZ	Exclusive Economic Zone
EFZ	Exclusive fishery Zone
GESAMP	Group of Experts on the Scientific Aspects of Marine Pollution (UN system)
ICSU	International Council of Scientific Unions
IFIAS	International Federation of Institutes for Advanced Studies
IGBP	International Geosphere-Biosphere Programme, UNESCO
IOC	Intergovernmental Oceanographic Commission, UNESCO
ISSC	International Social Science Council, UNESCO
m	metre
MARPOL	Convention for Prevention of Pollution of the Sea from Ships
nm	nautical mile
OALOS	Office for Ocean Affairs and the Law of the Sea, United Nations
OAPEC	Organization of Arab Petroleum Exporting Countries
OM	Ocean Management
OAM	Ocean Area Management
OETB	Ocean Economics and Technology Branch, United Nations
OTEC	Ocean Thermal Electric Plant
sq	square
UN	United Nations
UNCLOS	United Nation Conference on the Law of the Sea
UNEP	United Nations Environment Programme

CONTENTS

CONTENTS   ix

CONTENTS

CHAPTER 1

THE THEORETICAL CONTEXT

1.1 THE SCIENTIFIC IMPORTANCE OF SEA MANAGEMENT

Sea management has acquired scientific importance since the early 1970s [Kenchington, 1990, 6-13]. The first international event which provided stimulating circumstances was the Third United Nations Conference on the Law of the Sea (UNCLOS III) commencing in 1973. Discussions and proposals which were set up in that context, as well strategies encouraged, have both influenced national policies concerned with the sea and have given impetus to an international debate about the involvement of the ocean in human activities and in the exploitation of the earth's resources [Interfutures, 1979, 115-6; Mann Borgese, 1986, 13-41]. Because of these policies and strategies one of the strongest reactions in the relationships between society and nature in recent history has taken place (Figure 1.1):

(i)   research, which has been implemented by a singular advance in technologies, has provided wider and wider knowledge of physical processes and marine ecosystems, opening up new perspectives in the role of sea uses in both international and national environments;

(ii) political and economic strategies, aiming at securing and exploiting new resources for states, have benefited from these advances;

(iii) research progress has benefited from this political and economic interest;

(iv) progress in research has provided advantages for both the political and managerial worlds.

As a result, a circular relationship has been initiated and developed.

These circumstances make it evident that social objectives have had a leading role in giving impetus to the management of the sea: these have been an independent variable whereas research has been a dependent one. The consequences have been important: the ways in which practitioners look at the sea and ask for research derive from the fact that its management is first a political issue and then a scientific one.

1

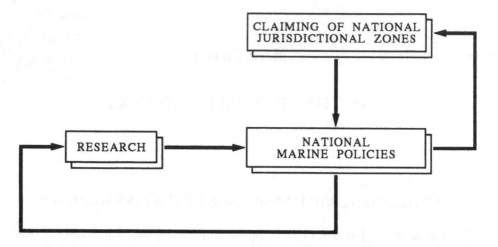

Figure 1.1 *Feed-back relations influencing sea management.* On the one hand, research and marine policy have largely interacted giving impetus to each other; on the other hand, in the context of the Third UN Conference on the Law of the Sea (UNCLOS III), circular relationships developed between the implementation of national marine policies and the claiming of jurisdictional zones.

In the political context three inputs had a leading role: (i) the changes in relationships between developed and developing countries; (ii) conflicts focused upon the international markets for raw materials and fisheries; (iii) environmental concern.

From the second half of 1960s until the late 1970s the de-colonialization process generated considerable influences in the United Nations system because the number of developing countries represented in the General Assembly grew to the point of influencing the search for a new international division of labour. To some extent the need to formulate a new Law of the Sea seemed to mirror that of the establishment of the so-called "new international order": according to a number of politicians and scientists, the former would have been a significant part of the latter [Mann Borgese, 1986, Chapter 6]. This willingness brought about both the setting up of 200 nm national jurisdictional zones, which have diffused throughout the ocean world, and the pursuit of the development of north-south co-operation in the exploitation of deep seabed resources. These historical troubles are too well known to be recounted here [Prescott, 1985, 81-107; Pardo, 1978].

The prospect of widespread exploitation of deep-sea nodule deposits, which was opened up by oceanographic research in the 1960s, and strengthened in the following decade, together with pushing the exploration of evaporite- and mud-endowed ocean areas, generated the dream that deepsea mining was, even in the short or medium term, the new frontier of the natural resource economy [Post, 1983, 75-87]. This dream has had two aspects: first, it was thought that the prices of minerals from land would have grown so rapidly so as to make it convenient to get these from the sea; secondly, the possibility of setting up technologies capable of exploiting the sea without significant environmental impacts was over-estimated. *Inter alia*, because of these prospects developing countries' policies were based on the certainty that deepsea mining would have been a powerful tool—maybe the sole one—in creating a different balance between developed and developing countries.

Meanwhile, the prospect of producing energy from the sea has given a great impulse to the creation of a promising atmosphere for marine resource exploitation. A number of technologies have been provided for the exploitation of both the physical properties and movements of the water mass and winds. A primary interest has been ocean thermal energy conversion, where appropriate technologies have been developed suitable for use in developing countries, where large quantities of energy could theoretically be produced [UN, DIESA, 1984]. However, such technologies were to some extent associated with unacceptable environmental impacts so no significant developments have been taken place to date. Other energy production patterns have also been criticised from the ecological point of view.

The most important advances were achieved in oil and gas exploration and exploitation. Apart from the role that this marine resource use has had for the evolution of the international energy system, its technology acquired special significance because it (i) has given strong impetus to the investigation of the seabed and subsoil of the continental margin, (ii) has encouraged the development of sophisticated systems to protect the environment and (iii) has established positive interactions with other sea uses, such as undersea archaeology and research.

In fact, in spite of being recent, environmental concerns have acquired influence so rapidly as to considerably affect economic strategies for the exploitation of both mineral and energy resources. As a result, these prospects depend to an ever greater degree on both the employment of technologies and organisational patterns, which seem to be no more expensive than those on land, and on the possibility of preventing damage to ecosystems and their physical environment. As a result, the exploitation of non-living marine resources has become much more complicated than was expected in the 1960s and 1970s.

As far as the exploitation of living resources is concerned, it is well known that fishery activities have grown and spread out over the oceans to the point of fundamentally affecting ecosystems in some areas, both in the intertropical and subpolar latitudes, and the catch of some species is so great as to risk destroying them. As a consequence, ecological concerns have become as fundamental as in the exploitation of non-living resources.

These crucial problems should be evaluated in the context of events at UNCLOS III during the 1970s: the prospect of paving the way for the exploitation of the sea gave impetus to the establishment of national jurisdictional zones including the enlargement of the territorial sea and the creation of the Exclusive Economic Zone. At that time the main goal consisted in providing a legal basis for the development of sea uses through an unusually rapid implementation of national jurisdictional zones. Although the 1982 UNCLOS includes a very advanced code of conduct for environmental management, the strongest stimuli were given to resource management.

These circumstances are worth recalling because they have deeply influenced the management of the sea. In its initial phase sea management, although it is a multidisciplinary issue, was approached through methodologies in which the law has played a leading role: managing the sea has consisted, first of all, in claiming or agreeing jurisdictional zones. Research on the environment and the setting up of scenarios of the relationships between resource development and environmental management have been regarded as subsequent tasks. As a consequence, the need to overcome this approach and to set up appropriate multidisciplinary-based methodologies is one of the primary tasks of the present time. This does not mean that maritime management should only benefit from the contributions from a large number of disciplines but also that these sectoral approaches should converge into the development of an holistic view.

Subject 1.1

Sea and ocean: definitions

According to the *Encyclopaedia Britannica*, the words "sea" and "ocean" are synonymous: both of them mean "the great body of salt water covering the larger portion of the earth's surface". As is well known, in literature on management two sets of meanings have been attributed to these terms. According to the first semantic approach, the marine water mass as a whole is called "sea", and in this context, (i) "sea" means also the marine mass in a abstract sense (e.g., sea uses, sea environment); (ii) the coastal area is usually—or tends to be—referred to the water column superadjacent to the continental margin; (iii) the ocean is regarded as the water column superadjacent to the ocean bottom beyond the outer edge of the continental margin. As a result, the language of management is founded on three categories: sea, coastal area and ocean area. According to the second semantic approach, the salt water mass of the planet as a whole is called "ocean". In this context (i) "sea" is used in an abstract sense, but "ocean" is used synonymously, (ii) the water column superadjacent to the continental margin is called "coastal", (iii) the water column beyond the outer edge of the margin is called "ocean". As a result, "ocean" has two meanings, since it refers to both the marine mass as a whole and the mass superadjacent the bottom of the ocean. In this book the first option is followed.

Analysis of sea management has been mostly empirical. The need for investigations, the objectives of which were to determine maritime boundaries, give form to national claims and deal with conflicts among sea uses, has brought about many analyses of specific cases. As a result, the development of theoretical and empirical research fields respectively have been out of balance, the former having had less attention than it requires, although since the late 1970s some relevant theoretical contributions have been produced. A significant example of these was the 1977 Progress Report of the United Nations [UN, Secretary General, 1977, Document E/5971]. Starting from a structuralist background it developed reasoning on integrated coastal area development and made some contributions relevant to distinguishing coastal from ocean management. But not many efforts of this kind have appeared so research has largely remained empirically oriented.

Subject 1.2
The birth of sea management

"During the decade of the seventies there was general recognition of the importance of marine resources for the economic growth of states, and increase in scientific research activities, and for a sustained negotiation effort at the international level that culminated in the adoption of the UN Convention on the Law of the Sea. Growing importance was attached to the opportunities that resource exploitation could offer within the context and objectives of national economic and social development. Concern of developed countries over the quality of their coastal and marine environments prompted enactments such as the U.S. Coastal Zone Management Act of 1972, and some specific measures in various European countries. These experiences, although not formally replicated in the developing world, had a profound impact there and later became the basis for concepts and approaches developed to tackle the problems encountered in the coastal areas of developing countries." [Vallejo, 1988, 205-6].

1.2 FROM THE COASTAL AREA TO THE OCEAN

The second event by which sea management was influenced was the US Coastal Zone Management (CZM) Act. Its influence has been so strong and has diffused so rapidly as to justify stating that this law, which came into force in 1972, was the real birth of sea management. Also thanks to this national legal framework, three issues have acquired scientific and practical importance: the delimitation of the coastal zone, conflicts among coastal uses and users and the formulation of management criteria.

Since coastal zone management has become one of the most complex tasks of coastal states the criteria for delimiting coastal areas have acquired more and more importance. These will be discussed in Chapter 7. Here it suffices to note that the setting up of such criteria brought into play the need to distinguish coastal from ocean management. As Vallejo points out [1988, 206-7], these two categories—Coastal Area Management (CAM), also called Coastal Zone Management (CZM), and Ocean Management (OM), also called Deepsea Management (DSM)—have become a sort of dichotomy. This distinction, which is provided with both pragmatic and scientific justification, would not have been so widely debated in the literature if the US Coastal Zone Management Act had not been passed.

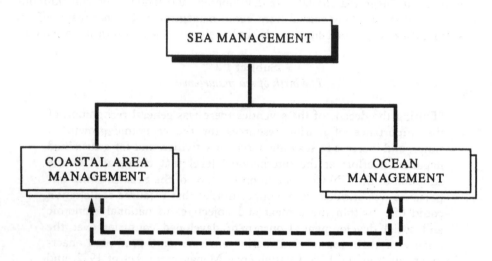

Figure 1.2 *The basic framework of sea management* consists of coastal area management and ocean management. The former was effectively born in 1972 (USA, Coastal Zone Management Act); according to Vallejo (1988, 208-9), the initial inputs towards the latter date back to 1984 (UNEP, Blue Plan).

Both the growth of sea uses and their diffusion in developed and developing countries have given strong impulses to conflicts between uses. New kinds of sea uses, such as the offshore oil and gas industry and undersea archaeology, have brought about new kinds of conflicts at sea. Users have become more and more numerous, while in many parts of the world institutional and corporate interests at sea have come into conflict. The need to put a restraint on uses risking the environment and to promote uses of cultural value, such as marine parks and reserves, has spread and acquired social acceptance. As a result, since the CZM Act entered into force national laws have had to deal with this concern, formulating codes of conduct for preventing and settling conflicts. First this concern involved practitioners, followed by researchers, who have been asked to set up methodologies to settle relationships between uses. It did not take long to realize that this concern was only the tip of an iceberg, the mass of which consists of the need to set up comprehensive management criteria [Armstrong and Ryner, 1981, 81-117; Mitchell, 1982].

In conclusion, sea management is evolving in three directions: (i) from the single or a few use-based to the multiple use-based patterns (growing complexity); (ii) from the shoreline sea areas increasingly further offshore (extending management space); (iii) from internal waters and the territorial sea to the Exclusive Economic Zone (jurisdictional and legal complexity).

(i)   *Uses.* Management patterns have been developed to tackle a growing number of uses co-existing in a given marine area. In its embryonic stage sea management dealt with single or a few uses—such as navigation, fisheries, and recreational uses—but recently, particularly from the mid-1980s, multiple use-oriented management patterns have been diffused. By virtue of this diffusion process, criteria to manage the sea should be developed to tackle more and more complicated networks of uses. This is one of the most stimulating concerns for research which is needed to provide appropriate planning-oriented methodologies.

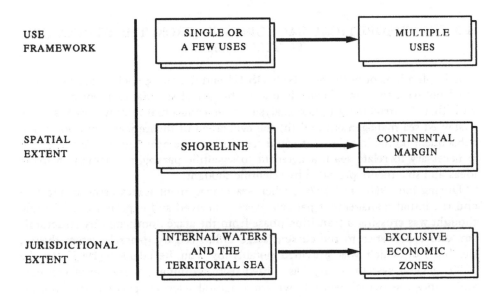

Figure 1.3 *The evolution of sea management.*

(ii) *Spatial extent.* Through the impulses to exploit coastal areas, management has involved larger and larger marine zones extending seawards. In the past, when coastal states were focusing their attention on shorelines, geomorphological features were the physical reference. Management, concerned for example with port facilities and recreational uses, was related to a narrow zone in which geomorphological features were suitable for building structures, such as breakwaters and offshore oil and gas terminals. In short, the technological level and the physical context of the land-sea interface were playing the leading role in management.

(iii) *Jurisdictional zones*. In order to provide new advantages and opportunities for coastal, island and archipelagic states four legal developments were made in the context of the UNCLOS III: criteria to establish the baselines were modified with the aim of enlarging the extent of internal waters; the extent of the territorial sea was established at 12 nm from the baselines; the continental shelf was maintained but it was conceived in a different way from the past; and the Exclusive Economic Zone was introduced. Hence two consequences: coastal area management was thought of as covering the widest jurisdictional zones claimed or agreed by the coastal, island or archipelagic states; and the ocean area was regarded as consisting of a space subject to the international régime.

1.3 THE THEORETICAL BACKGROUND: FROM THE STRUCTURE TO THE SYSTEM

The implications of both the UNCLOS III and national policies have been outlined not with the aim of considering the history of sea management *per se*, but with that of introducing (i) the analysis of the conceptual frameworks that have sustained sea management and (ii) the evaluation of theoretical frameworks and subsequent methodologies that could sustain it in the near future. To this end it is necessary to relate sea management to scientific paradigms, namely, to basic ideas and statements provided by scientific thought.

During the 1960s and 1970s, when sea management issues came to the fore and the initial management patterns were conceived and experienced, scientific thought was crossing a transition phase from the stage dominated by structuralism and cybernetics to the current stage influenced by general system theory, the basis of which was provided notably by Von Bertalanffy [1968]. In the 1980s general system theory has shifted from its initial generation of concepts and principles—which were drawn from natural sciences, particularly from biology—to a new stage in which social sciences, including sociology, have exercised considerable influences. This was mostly due to the theory of complexity which has brought into evidence the need to provide a general system approach with a clear non deterministic background [Le Moigne, 1985; Morin, 1985]. Bearing in mind the turning point from structuralist thought to general system-based principles in the 1970s and 1980s it seems appropriate to wonder: (i) to what extent sea management has been supported by structuralism; (ii) what implications could arise if it is supported by concepts and principles provided by general system theory. In other words, the question is what advantages sea management patterns could derive from the adoption of criteria consistent with the most advanced general system theory—and theory of complexity, as well (Figure 1.4).

Subject 1.3
The theory of general system and the general theory of systems

As is well known, a part of the literature about this field states that the term "general" in "general system theory"—the expression by Van Bertalanffy (1968)—is to be referred to "system" and, as a result, the speculation should achieve the setting up of the *theory of the general system*. In this case the "system" is regarded as a general model (Le Moigne, 1984, 284). On the other hand, a number of epistemologists state that the term "general" refers to the "theory", namely, to its nature. As a result, the objective of the speculation is the setting up of a *general theory of systems*. The conflict between these approaches is discussed by Le Moigne (1984) in a book with a significant title: *La théorie générale du système* (283-4). In the present analysis about the management of the sea the former approach is followed.

In this context it should be borne in mind that, according to the structuralist paradigm, reality is regarded as a set of elements tied by a set of relationships and evolving over time. This paradigm has provided the conceptual background for sea management—and for the ways of looking at the sea as a whole—up the present time. Such an involvement appears self-evident when it is considered that the totality of sea uses is usually represented by square matrices—which will be discussed in Chapter 5—in which sea uses are classified with the aim of paving the way to multiple use-based management and of settling conflicts between uses. These matrices, which have this general form:

SEA USES X_j

SEA USES X_i RELATIONSHIPS X_i/X_j

and are commonly called "global marine interaction models" [Couper ed., 1983, 208–209], are the formalized representation of the sea structure and have provided a considerable impetus to rationalise research on sea planning and management. In fact they are use-use relationships models.

By the structuralism-based paradigm reality is imagined as a set of structures generating functions and evolving [Le Moigne, 1984, 50–54; Vallega 1990/b, 37–60]. Evolution is thought of *as only the movement* of the structure over time. The *change of the structure* during its evolution, as well the nature of change and the possibility that change leads to the disappearance of the structure and the establishment of another, are not issues consistent with this approach. Apart from the subsequent theoretical implications, it is worth bearing in mind that, as far as sea management is concerned, two consequences have arisen.

(i)   The structuralist approach has not provided fertile ground for taking deeply into consideration the relationships between sea uses—through which behaviour towards the sea reveals itself—and the marine eco-system. This has been due to the fact that the subsequent theoretical background of sea management has not investigated in which ways the ecosystem can collapse—i.e. it was not concerned with the morphogene-sis of the sea structure. *Optimo iure*, human behaviour patterns leading the ecosystem to collapse were not the foci of structuralist analysis.

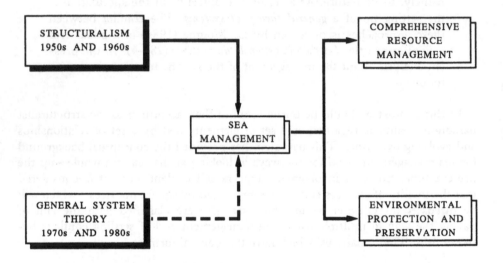

Figure 1.4 *The theoretical and methodological support of sea management*. While sea man-agement has evolved towards the integration between resource use and environmental protection and preservation, it has been, however, influenced by structuralism. At the present time the question is how general system theory can support it.

(ii) This approach has not been consistent with the literature aiming at inves-tigating global change—also involving the ocean world—just because it did not face the ecological implications of sea uses as the main research issue. As a consequence, a gap between the thought supporting research on global change and structuralist approach has come to light. The main reason was due to the fact that research on global change is sustained by a point of view which is antithetic as regards paradigms: radical changes in reality, consisting of morphogeneses is considered as normal, i.e. consis-tent with the nature of the world.

Because of these circumstances attention tends to shift from structuralism to system-based thought with the aim of evaluating what implications are engen-dered if the latter conceptual framework sustains sea management. As is well known, the general system-grounded metaphor is more complex than the structuralist one: reality is not only conceived as a structure generating functions and evolving, but it is also referred to the external environment and the objectives towards which it moves. Three new theoretical inputs arise: the external environment, the change of the structure and the objectives of the structure. In particular, theoretical complexity is due to both the external environment and teleology: (i) structure can be identified and investigated only if the interplay between its elements and the external environment are sufficiently taken into account; (ii) the evolution of structure cannot be perceived and described if the goals towards which the structure is moving are not the foci of analysis.

As a result, following a general system theory-inspired approach, the first objective of the analysis consists of the identification of the sea structure and its interaction with the external environment. The sea structure has two components: the natural one, i.e. the ecosystem and its physical context; and the social one, i.e. sea uses.

At the methodological level the main contribution provided by the system-oriented approach is the employment of models as tools to describe and explain reality. According to Le Moigne [1984, 269–70], this implies passing through three methodological stages: (i) the creation of the model, (ii) the implementation of the model and (iii) the simulation of sea management patterns through the model. Applying these system theory-inspired methodological principles, sea management analysis could develop through three stages.

(i) *The setting up of the model.* While the structuralist approach starts from the elements of the structure and their reciprocal relationships, the starting point of the system-based approach is the objective to which the sea structure moves because of resource management and subsequent environmental implications. In the subsequent step the role of the external environment is taken into account with the aim of realising which changes the sea structure undergoes because of the goals towards which it moves and the inputs that it receives. On this basis, consideration is centred on (i) the changes of the functions of the structure and (ii) the evolution brought about by the change of functions. Because the evolution generates changes in the goals of the sea structure, the model of the sea structure tends to appears as a circular relationship involving goals, external environment, structure, functions, and evolution (Figure 1.5, section I). Of course, the consideration of structural evolution leads to that of the goals of the sea structure, so the analysis continues following a circular path. This logical process mirrors what happens when a management plan concerning a coastal or an ocean area is built up

(ii) *The implementation of the model.* In this phase the task of analysis changes: since the model exists, what should be done is to adjust it with the aim of making it consistent with reality, i.e. the evolution of sea uses and environmental impacts. This time the functions of the sea structure are the starting point of analysis: they are considered in order to know their evolution—the general system-based analysis being dynamic in its nature. In this context it is possible to understand whether and to what extent the evolution of the sea structure—and the subsequent evolution of the relationships between the structure and its external environment— bring about the goals of the structure itself. In conclusion, when the model is implemented the sequence of steps is different from that which takes place in the setting up of the model. This time the sequence concerns functions, evolution, external environment, objectives and structure. It may be pointed out that the analysis of the changes in the sea structure leads to that of the functions of the structure: as a consequence, the implementation of the model appears again as an endless circular process (Figure 1.5, section II).

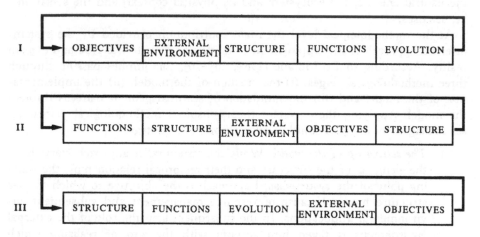

Figure 1.5 *The logical processes grounded on general system theory*: I, the setting up of the model; II, the implementation of the model; III, the simulation of the structure. From Le Moigne, 1985, 270, with adjustments.

(iii) *The simulation of the sea structure.* As is well known, in management-oriented research, models are fruitfully employed also with the aim of realising what can occur when given management patterns are applied. This use of models is proper to sea management especially when it acquires sophisticated features, as commonly takes place in coastal areas. The simulation should start from the consideration of the structure and proceed to evaluating which implications can take place by its changes. As a consequence, the evolution of the use structure is considered focusing attention upon the interaction of the structure with the external environment and on the subsequent changes of goals of sea uses and environmental management. It is self-evident that the sequence of steps of the analysis is different from either the creation and implementation of the model: this time it consists of structure, functions, evolution, external environment, objectives (Figure 1.5, section III).

1.4 THE GENERAL SYSTEM-BASED APPROACH

The employment of models, which act as metaphors of reality, and the development of the dynamic approach, are the methodological watershed between structuralism and general system theory. Because of that it is opportune that the role of concepts and isomorphisms provided by the general system-based approach are taken concisely into account.

Sea structure. As far as management is concerned, the sea consists of a set of structures. Each of them is more or less directly or indirectly involved in the human presence and activities. According to current scientific thought, a structure is a set of elements reciprocally tied by relationships that are closer than those through which the structure interacts with its external environment. Two features of the sea structure are worth considering: its nature and its extent.

As far as its *nature* is concerned, the structure reveals itself as complicated because it consists of two components (Figure 1.6): (i) a natural component; (ii) a social component. The natural component consists of an ecosystem and its physical environment, the nature of which will be discussed in the next Chapter. The social component consists of the human presence in the marine environment and consequent sea uses. From the conceptual point of view, this structure mirrors the specific features of each structure in which man interacts with the natural environment and, in particular, includes three sets of relationships taking place between:

(i)   sea uses (i.e. between settlements, plants, carriers and activities at sea);

(ii) the elements of the ecosystem and its physical environment (i.e. between geological, geomorphological, chemical and biological elements).

(iii) the sea uses and the ecosystem.

Most relationships are complicated because they consist of series, parallel, feedback and multiple compound chains of links [Harvey, 1969, 454]. In particular, the feedback relations between sea uses and the natural environment are noteworthy, continually acquiring greater importance.

This structure will be called *sea use structure* to stress its importance for management. Its *extent* is due to the spatial location of its elements and the subsequent area covered by the web of relationships among elements. From the methodological point of view the identification of this extent is one of the most interesting concerns: regional theory, which is the disciplinary field dealing with this subject, has still not provided satisfactory tools. In any case, if management is the key objective of analysis, the sea use structure has to be identified and its extent has to be delimited according to the goals to be pursued. In this context,

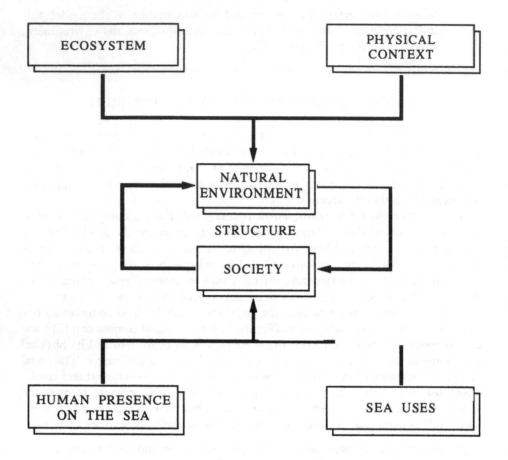

Figure 1.6 *The sea use structure* consisting of two components, natural and social

research is goal-oriented, as the general system-inspired approach requires, and two spatial categories come to light: coastal and oceanic. These can be distinguished using both physical and legal criteria, or a combination of them; this issue will be specifically considered in Chapters 7 and 9.

External environment. The external environment also consists of two components, natural and social, which interact with each other (Figure 1.8). As is well known, the inclusion of the external environment in analysis—particularly when it is management-oriented—is one of the most innovatory features of general system theory *vis-à-vis* structuralism. The natural component of the external environment embraces a large range of processes, such as plate tectonics, climatic change, ocean circulation, and biological macro-evolution. The social component consists of national policies, international co-operation and conflicts, the economic strategies of companies, technological advance, and social perception of the sea.

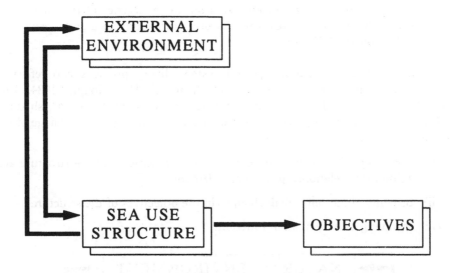

Figure 1.7 *The evolution of the sea use structure* provided by the general system-based approach.

Process and change. The structuralist approach deals with the evolution of the structure, while the general system theory approach deals with the idea of process and—what is much more important—joins together the ideas of process and change. The concept of the structure moving towards goals and the changes which it undergoes during its evolution are other aspects differentiating the system-based approach from the structuralist-inspired one. This has special importance for sea management because, when the evolution of sea use structure is investigated with the aim of ascertaining what change it is undergoing or can undergo, it is possible to set up strategies and technologies to maximize advantages and minimize environmental impacts.

Subject 1.4

Key concepts of the theory of complexity

As far as the perspectives opened by the theory of complexity are concerned, "it seems possible to start (...) drawing upon key concepts used and key issues addressed. The following would be included in such a sample. First, key issues are reflected in a loose cluster of principles and emerging paradigms that are defined by key concepts such as the interrelationships between order and disorder, the creation of over-increasing complex orders out of noise (...): autopoiesis, self-regulation, and spontaneous self-organization in natural and social systems (...). Inherent in these concepts is a reaction against determinism, a new acceptance of instability, chance, possibility, and of

stochastic processes, a new emphasis on the emergence of the unexpected, the novel, the creative—and of a new significance and meaning". [Ploman E. W., 1985, 9].

In its simplest formulation, general system theory provides two reference
concepts of change: adjustment and morphogenesis [Le Moigne, 1984, 197-
211]. It is not easy to define these and, *optimo iure*, to establish the watershed between these. As a first approach, adjustment can be thought as any change of the
sea use structure:

(i)   consisting of the change of only a subset of elements of the structure and
      subsequent relationships between elements;

(ii) bringing about only small changes in the objectives of the structure;

(iii) involving only regulation functions.

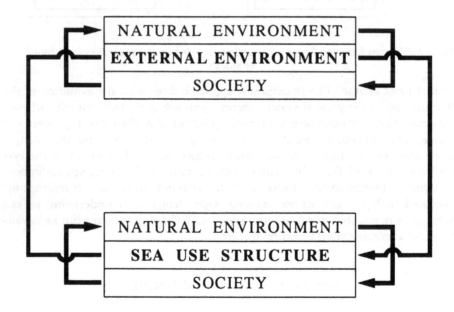

Figure 1.8 *The interaction between the sea use structure and its external environment.*

On the other hand, morphogenesis can be regarded as a change of the sea use
structure:

(i)   consisting of the disappearance of all or most elements of the structure
      and subsequent relationships between elements;

(ii) bringing about radical changes in the objectives of the structure;

(iii) to the point of giving birth to a new structure.

Both adjustment and morphogenesis depend on endogenous factors, acting in the context of the sea structure and/or exogenous factors, acting in the external environment. Bearing that in mind, this general scheme from Le Moigne [1984, 197] can be regarded as a reference point of reasoning about the evolution of the sea structure, i.e. of the coastal and ocean areas.

TABLE 1.1
Evolution in the structure and external environment

| | | objectives | |
		stable	changing
	stable	regulation (adjustment)	adaptation (adjustment)
external environment			
	changing	adaptation (adjustment)	transformation (morphogenesis)

As a tentative approach, the interaction of mechanisms acting both inside the sea structure and in the external environment include four main cases.

(i)   Both the external environment and objectives of the structure do not change. In this case the structure undergoes a simple regulation process.

(ii)   The external environment does not change while the objectives of the structure do. In this case the structure undergoes greater or lesser adjustments and maintains its features.

(iii) The external environment changes while the objectives of the structure do not. In this case adjustments, consisting of the establishment of new kinds of mechanisms, take place in the structure.

(iv)   Both the external environment and objectives of the structure change, provoking a real morphogenesis of the structure. In this case the webs of the elements of the structure radically change, new sets of relationships take shape and, as a consequence, the structure moves towards new goals.

The most interesting case occurs when impulses from the external environment change and the action of endogenous factors draw together carrying the structure to new goals resulting in morphogenesis: at the present time many marine areas undergo this kind of process. As a consequence, the analysis of morphogenetic phases is presently very rewarding, particularly in the view of implementing the rationale of sea management.

Bifurcations in the evolution of sea structure. Because of this relevance, what happens in the context of the sea use structure when it enters a morphogenetic phase is one of the most fascinating issues raised by general system theory. Recently Laszlo [1988] stated that this phase can be explained by the macro-evolution theory because it is an isomorphism capable of explaining changes both in ecosystems and societies. According to the macro-evolution theory, during its path over time the structure is supposed to cross phases in which alternative directions arise. In these phases the structure must choose between alternative futures: it faces a bifurcation. The idea of bifurcation, which implies that of choice, presupposes that reality—regarded as a natural and social system—is not conceived in a deterministic way. This statement reverses the common way of thinking about the sea use structure which considers the ecosystem and its physical context in a deterministic way—and the human communities in a non-deterministic way, of course.

Sea use structure as a non-trivial machine. The idea that not only society but also the ecosystem act in an indeterministic way is innovatory in sea management. How revolutionary this innovation is, both in scientific approach and in subsequent management patterns, may be understood by introducing the concepts of trivial and non-trivial machines.

According to Von Foester [1985, 127-33], a structure which acts deterministically is a trivial machine: in this case each input that the sea structure receives always brings about the same output. This is due to the fact that the machine is not able to change its internal state. On the other hand, a structure acting as a non-trivial machine is able to change its internal state independently of the input to which it is subjected. As a consequence, a range of outputs is related to one input. In the context of sea management, the innovatory approach concerns the ecosystem.

When the ecosystem is regarded as a trivial machine—i.e. according to the conventional structuralist approach—it is thought of as always reacting in the same way to the same impulse from human presence on the sea and sea uses. In other words, when the input A is directed from human presence and activity at sea to the marine ecosystem and its physical context, the ecosystem always responds with output A'. As a consequence, it is possible to predict which kind of responses the ecosystem will give to a given management pattern: the future is thought of in a deterministic way, as the *product of the past.* This a most important implication for sea management: it is reassuring for both researchers and managers because it induces the conviction that the coastal and ocean areas can always be under control on account of their mechanistic behaviour. This also has an ethical implication: societies are certain to reach any objectives in sea management provided that they are able to set up techniques to prevent—or to reasonably restrict—the negative environmental impacts that they *have imagined* as the effects of sea uses. Determinism provides certainty in knowledge, and certainty in knowledge encourages the implementation of sea uses.

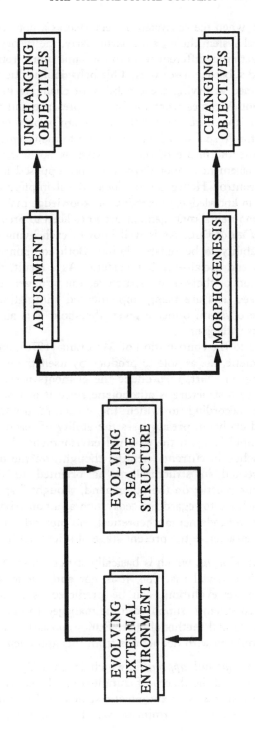

Figure 1.9 *The change in the sea use structure.*

On the contrary, when the ecosystem is regarded as a non-trivial machine—i.e. according to general system theory and, particularly, the theory of complexity—it is thought of as reacting in different ways to the same input from the human presence on the sea and subsequent sea uses. This behaviour is due to the fact that the ecosystem is imagined as having the capability of changing its internal state. As a result, when input A comes from the human presence and activity at sea, the marine ecosystem is able to respond with a spectrum of outputs: A', A'', A''', etc. In this case the future responses of the ecosystem to management patterns cannot be predicted because the future of the ecosystem is not a direct product of the past—and sea management cannot anyway be presupposed as able to keep the ecosystem under control. Hence the obvious ethical implication: indeterminism causes uncertainty in knowledge, uncertainty in knowledge makes sea management cautious, and environmental management becomes keen to preserve ecosystems.

Homeostasis and autopoiesis. As is well known, each living system is homeostatic: it has the ability to be self-regulating. Both components of the sea use structure—natural and social—are homeostatic. As a result, the higher the organisational cohesion of the sea use structure, the stronger its homeostasis. At the present time sea uses are being implemented stimulating the homeostatic behaviour of the ecosystem: under a given threshold it is able to self-regulate; beyond that it risks collapsing.

Recently, thanks to the contribution of Maturana [1978], any living system is regarded as autopoietic, i.e. as able to produce by itself the elements for surviving and developing: in short, to produce the elements necessary to sustain its evolution. Of course a structure is autopoietic since it is a non-trivial machine: this circumstance, according to which the structure produces elements for adapting itself and evolving, presupposes the ability of reacting in a non-deterministic way to impulses from the external environment. As a result, a watershed has been reached in current scientific thought: on the one hand, conventional thought, rooted in structuralism, is oriented to thinking that the ecosystem is not autopoietic; on the other hand, thought inspired by the theory of complexity is inclined to regard the ecosystem as a non-trivial machine.

In conclusion, two fundamental theoretical and methodological approaches to sea management characterize the present stage of scientific thought.

(i) *Conventional thought,* which is basically structuralist. According to it, the ecosystem is a trivial machine, it is not autopoietic and its future behaviour—i.e. its evolution—can be predicted as a consequence of the past. The ecosystem is supposed to be managed through mechanistic theories, concepts and methodologies. Ethics consists of behaving in such a way as to conform with deductions from this approach.

(ii) The *non-conventional approach,* which is basically inspired by general system theory and the theory of complexity. The ecosystem is a non-trivial machine and acts as an autopoietic structure. Its future behaviour cannot be predicted in a deterministic way because it is not regarded as the consequence of the past. Ethics consists of minimizing any implications on the marine ecosystem of the human presence and activities at sea. This statement might be regarded as radical and unjustified but it could be a stimulating basis for discussing the role of the human presence at sea.

1.5 EXPLAINING MANAGEMENT: TWO CRITERIA

On the basis of the general system and the complexity-based approach attention may be centred on the nature of changes which the sea use structure can undergo. At the general level, Le Moigne [1984, 236-51] suggested explaining the evolution of structures through the differentiation-coordination model. The objective of this methodological tool consists of identifying when, in spite of being differentiated, structural elements are coordinated enough to maintain appropriate evolution (Figure 1.10). Transferring this model to sea management a Cartesian diagram can be imagined: it is based on the coordination axis (X) and the differentiation one (Y).

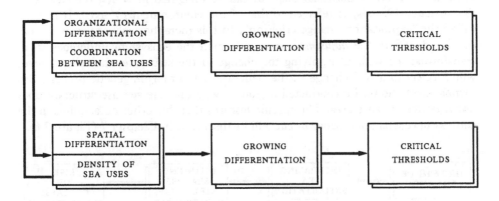

Figure 1.10 *The differentiation-coordination criterion*: the conceptual basis.

As the number of sea uses grows the sea use structure undergoes two kinds of differentiation. First, *organisational differentiation*, which is due to the fact that more and more resources are exploited and the marine environment is influenced by a growing number of facilities and activities. Secondly, *spatial differentiation*, which is due to the circumstance that the marine area is involved in different kinds of resource uses. While differentiation advances, the need for coordination between sea uses, i.e. activities and facilities at sea, also grows. When co-ordination between these slows down below a given threshold, the sea use structure is jeopardized. As a result, it is necessary that sea management advances in such a way that the differentiation of the structure does not cause an unbearable failure of co-ordination. The range of differentiation-coordination relationships can be represented in our Cartesian diagram and sea management can be supposed to be asked to prevent differentiation pushing forward, and co-ordination slowing down below a critical threshold. The Cartesian diagram referred to coastal management will be discussed in Chapter 7 (Figure 7.5).

The *differentiation-coordination criterion* is a tool to investigate the final result of resource management and is strictly concerned with the use framework. Moving to environmental management another complementary criterion may be built: the *exploitation-change criterion* (Figure 1.11). It is not considered in the literature but it seems to be a logical consequence of the reasoning developed until now and consists of a dynamic view of the relationships between sea uses, on the one hand, and the ecosystem and its physical context, on the other. This criterion derives from the consideration that the evolution of the sea use structure is influenced by the growth of sea uses. The final result is a change of the structure, which—as has been discussed—consists of adjustments or morphogeneses. Bearing in mind that one of the main objectives—or, at least, *the* objective—of sea management is the protection and preservation of the marine environment, analysis is required to place the change in relation to environmental impacts. So another Cartesian diagram can be imagined in which the change of the structure (X axis) is referred to that of the environment (Y axis). It may be called the "exploitation-change criterion". In this metaphorical context two areas can be identified: the *adjustment area*, in which the ecosystem undergoes small transformations without implying the change in the structure; the *collapse area*, which refers to a real change in the structure, i.e. a morphogenetic process. Sea management should be conducted in such a way that the sea use structure does not enter the collapse area. The specific features that this criterion acquires in the context of coastal area management will be discussed in Chapter 7 (Figure 7.6).

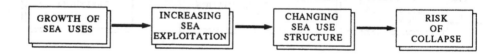

Figure 1.11 *The exploitation-change criterion*: the conceptual basis.

Both metaphors are epistemological, if it agreed that epistemology explains the nature of our experience in the world [Von Foester, 1985, 116]. As a result, they are bases for theoretical reasoning and the setting up of methodologies related to sea management.

1.6   CONCLUDING REMARKS: THE DILEMMA

The scientific approach to sea management deals with a kind of dilemma which is present in several other research and management fields. The dilemma concerns the transition from a structuralism-inspired to a general system theory-supported approach. As in the other fields, the structuralist approach has given some important advantages to sea management and related research: it has enabled researchers to cluster and investigate sea uses, coastal and ocean environments and decision-making centres to adopt sea management patterns. The need to overcome this approach has come to light because of three impulses: (i) the growth of sea uses has advanced so quickly as to cause important changes in management patterns of many coastal areas; (ii) the extent of the coastal area has enlarged bringing about relationships between uses and the natural context more complicated than those involved in the past; (iii) changes, due to the growth of sea uses and subsequent environmental impacts, have taken place in many ecosystems. The sum of effects caused by these factors highlights not only the need to implement dynamic analysis but also—and with special emphasis—to investigate the change phase and its nature. This is leading to a new approach, to some extent based on general system theory, and allows researchers to investigate how changes happen in a given context-such as that of the coastal area—affected by growing complexity. Theoretical analyses, which have taken place in the International Geosphere-Biosphere Programme (IGBP) and the Human Dimensions of Environmental Change Programme (HDECP), also encourage development in this direction. Such a theoretical and methodological passage implies: (i) the acquisition of new concepts and principles; (ii) the search for isomorphisms capable of describing contextually what occurs in marine exploitation and in the ecosystem.

Because of these inputs the analysis will first consider the evolution of the natural environment (Chapter 2) and the involvement of the sea in policies and strategies (Chapter 3). On this basis the interaction between the legal and natural environments (Chapter 4) will be taken into account with the aim of investigating sea management in its general features (Chapters 5 and 6). After that, coastal area (Chapters 7 and 8) and ocean area (Chapters 9 and 10) managements will be considered.

CHAPTER 2

THE EVOLUTION OF THE
NATURAL ENVIRONMENT

2.1 MARGINS AND OCEAN AREAS

Sea management deals with endless complexity. In order to consider it, attention should first be centred on the evolution of the natural environment beginning with the outer edge of the continental margin, linking the continents to the ocean basins. The margin is made up of continental crust and, except in some areas, consists of the shelf, slope and rise. The *continental shelf* extends from the shoreline to the so-called shelf break, beyond which the seafloor grades markedly to the slope acting as the interface between the adjacent continent and the sea. "The average depth of the shelf break is uniform, averaging about 130 m over most of the world ocean (...). The width of the shelf ranges from a few kilometers to greater than 400 km (...). The present level and gross topography of the continental shelf is the cumulative effect of erosion and sedimentation related to numerous large-scale sea-level oscillations during the last 1 million years" [Kennett, 1982, 29]. The *slope* grades from the shelf-break to a depth of 1500-3000 m and in many continental margins is divided by faults into more or less numerous escarpments. The *rise* is a province, with a gentle seaward gradient leading to the abyssal plains. Submarine canyons and deep-sea channels facilitate the transport of sediments to the deep ocean floor. *Trenches* can be very deep (from 2 to 11 km, and long)—the Peru-Chile trench is 5900 km—and tend to form systems.

TABLE 2.1
Area, volume and depth of the oceans

ocean and adjacent seas	area million sq km	volume million cu km	mean depth (m)
Pacific	181	714	3940
Atlantic	94	337	3575
Indian	74	284	3840
Arctic	12	14	1117
All oceans	361	1349	3729

From Kennett, 1982, 25.

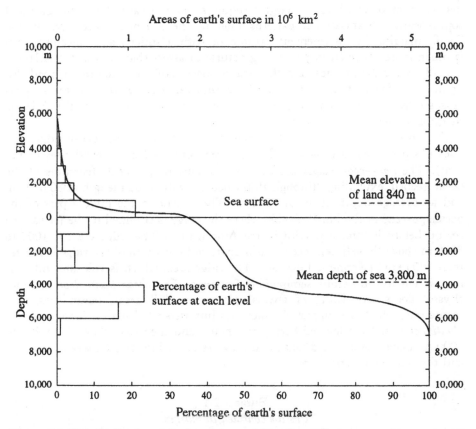

Figure 2.1 *The distribution of levels on the earth's surface.* The hypsographic curve indicates the percentages of the earth's surface that lie above, below or between any levels. From Kennett, 1982, 24.

Atlantic and Pacific types of margin influence sea management in different ways [IOC, 1984, 77-80]. As is well known, the *Atlantic margins* are made up of all the components of the continental margins—i.e. shelf, slope and rise—and are seismically inactive. By contrast, the *Pacific margins* consist of the shelf, slope and a deep trench and can be found in two sub-types: the *Chilean type*, which has a narrow shelf with a trench below the slope, and the *Marianas type*, which has a shallow marginal basin separating the continent from the island-arc and trench. Pacific margins are seismically active.

Hence three main consequences. First, the morphology of Atlantic continental margins, surrounding almost all the Atlantic, Arctic and Antarctic oceans, is less complicated than that of the Pacific margins, which mark the interfaces between the Pacific Ocean, the Americas and most part of Eastern Asia, the western side of Indonesia and other marine areas. As a result, in the Pacific the management of continental margins requires more advanced technologies and sophisticated organisational patterns than in the Atlantic. Secondly, on the Pacific margins sea management has to cope with seismicity and vulcanism so it is much more involved in preventing natural disasters than it is in the Atlantic. Thirdly, since the physical and chemical properties of water column depend also upon the effects of seismicity—such as mineralisation and evaporation processes-*ceteris paribus* marine ecosystems differ as we move from the Atlantic margins to those of the Pacific.

Deep ocean areas, covering 79 per cent of the ocean, embrace ocean ridges and rises (33 per cent), ocean-basin floor (41 per cent) and continental rises (5 per cent). The *mid-ocean ridges* are the most important physical structures of the ocean seabed, extending "through all the oceans, with a total length of 80,000 km and an average depth of about 2500 m. They occur in the middle part of the oceans, except in the North Pacific where the ridge is confined to the far eastern region before it intersects with North America (...) The ridge crest is 1000 to 3000 m above the adjacent ocean basin floor, and the width of the ridge is greater than 1000 km" [Kennett, 1982, 30]. The ridge area, which is volcanic and seismic, is cut by numerous semiparallel fractures. The *ocean-basin floor* includes abyssal floors, consisting of abyssal plains and hill areas, ocean areas and seamounts, which are almost all volcanoes but associated with ridge vulcanism. *Continental rises* are located between trenches and the ocean floor, from 200 to 1000 m above the seaward floor, and they act as an interface between the ocean and the continental structures.

Subject 2.1

Passive continental margins

"In many ways passive continental margins are among the most poorly understood areas in the oceanic crust. A principal reason for this is that many passive margins were formed by breakup a long geological time ago. The structures that were created initially and much of their subsequent history often lie buried under a thick cover of sediments.

The study of passive margins is very important for exploration of hydrocarbons, especially as the search moves into greater depths of burial and into deeper waters. It is obviously also very important for understanding the mechanisms operating near the time when continents separated to form new oceans." [IOC, 1984, 77]

During recent times, particularly since the 1960s, ocean areas have attracted attention because of strategies encouraging deepsea mining. Mineralisation has become an important topic of research. In the meanwhile technological advances in living resource exploitation has encouraged the development of ocean

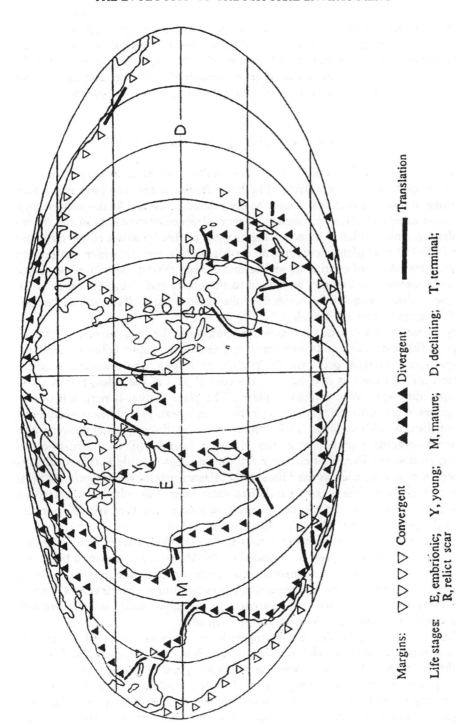

Margins: ▽▽▽ Convergent   ▲▲▲ Divergent   ▬▬▬ Translation

Life stages: E, embrionic; Y, young; M, mature; D, declining; T, terminal;
R, relict scar

Figure 2.2 *The distribution of margins and the life stages of oceanic spaces.* From Kennett, 1982, 322.

fisheries, increasing the need for research on ocean ecosystems and related food webs. As far as other sea uses are concerned, the ocean physical environment has had less importance than continental margins. But this gap could be reduced in the future because marine activities are moving towards ocean spaces and the need for more and more knowledge of ocean elements and processes is increasing.

2.2 EVOLVING CONTINENTAL MARGINS

In order to draw deductions from the analysis of natural environments a dynamic approach is unavoidable. The main reason is that the processes which continental margins undergo should be considered in detail. At the present time the most comprehensive explanation of natural processes consists of the theory of plate tectonics which is based on the analysis of the dynamics of the rigid and thin (100-150 km) plates covering the earth's surface and on which the present position and shape of continents and oceans depend. From the mid-ocean ridges hot volcanic materials exude, are added to the plates and shape the ocean crust. On the landward edge of the ocean floor the ocean crust is destroyed in the sub-duction zones, under the trenches.

By virtue of this crustal mechanism a margin can be (i) constructive and divergent, (ii) destructive and convergent, and (iii) conservative (Figure 2.2). *Constructive plate margins* occur on the mid-ocean ridges, where oceanic basalt is created and, through its motion from the central part of the ridge, brings about the spreading apart of the adjacent plates. The plate motion is perpendicular to the ridges. *Destructive margins* occur in trenches where two plates converge and one of them is pushed down. This process embraces three types of convergence: (i) ocean-to-ocean plates, such as the Marianas Islands; (ii) continent-to-ocean plates, such as the Peru-Chile trench and the adjacent cordillera; (iii) continent-to-continent plates, such as the Himalayas-Tibetan region. *Conservative margins* occur where two plates move parallel to each other but with different speed and/or in opposite directions. As a result, plates are neither created nor destroyed [Kennett, 1982, 320-7; Turekian, 1976, 114-8].

The present plate dynamics belongs to a process which started about 250 million years ago when Pangaea, surrounded by the Panthalassa, began to break up. According to the latest developments in plate tectonic theory, this process should end in 200-250 million years when reassembling of continents will occur and another super-continent, surrounded by a super-ocean, will be created. This implies that the *Earth's crust evolves cyclically*.

Plate dynamics provides the framework—a temporal scale of about 250 million years—within which each ocean completes its life cycle. As Kennett emphasises [1982, 179], each ocean goes through the following stages: embryonic, young, mature, declining, terminal, relict. During the embryonic and young stages uplift is the main motion; spreading and creation of mid-ocean ridges develop during maturity; compression and the creation of trenches mark the declining stage; lastly, compression and uplift, and the birth of mountain chains as well, occur during the terminal stage leading to the death of the ocean.

The features of the embryonic stage are manifest in the East African rift and those of the young stage in the Red Sea. Because of the creation of ridges, the Atlantic Ocean is going through its maturity stage. On the contrary, compression shows that the Pacific Ocean is declining: the impressive trench systems evidence this and how the ocean moves from maturity. The Mediterranean Sea, a heritage of the Tethys Ocean, has entered its terminal stage giving shape to the young mountain chains. The Himalayas and Indus area is a good example of what takes place when an ocean disappears.

During the ocean's life cycle margins evolve in different ways according to whether they are divergent or convergent.

Divergent margins move through three main phases.

(i)   *The rifting phase*, that at the present time occurs in the East African rift and marks the embryonic stage of an ocean.

(ii) *The onset of spreading phase*, in which the separation of the continental crust and the accretion of the oceanic crust between continental blocks come about and develop. This stage, at the present time affecting the Red Sea, implies a rapid regional subsidence of the outer shelf and slope.

(iii) *Maturity*, which is marked by slumping, canyons and deep-sea current erosion and deposition. Many parts of the present Atlantic and Indian margins are going through this stage.

Convergent margins bring about subduction, which is one of the most important effects of plate tectonics. Three consequences arise where one lithospheric structure is totally or partially inserted below another:

(i)   *consumption* of oceanic crust, caused by the downward displacement of deep-sea sediments through several kilometres beneath the convergent margin;

(ii) *scraping*, involving the skimming of sedimentary and igneous rocks from the upper part of the oceanic lithosphere;

(iii) *accretion* of the rock mass due to the materials added by scraping to the outer continental margin.

Margins involved in subduction give rise to two patterns: on the one hand, the *continental subduction-area pattern*, lacking a back-arc basin; on the other hand, the *island-arc subduction-area pattern*, including a back-arc basin.

Following this concise survey of physical elements influencing sea management, it is useful to summarise key points regarding the ocean bottom.

(i)   Divergent margins (i.e. the Atlantic type), which are passive and non seismic, are provided with a shelf width of up to 1500 km, a slope from 20 to 100 km, and an extended rise.

(ii) Convergent margins (i.e. the Pacific type), which are active and seismic, have a narrow shelf from 20 to 40 km, a steep slope and may have trenches.

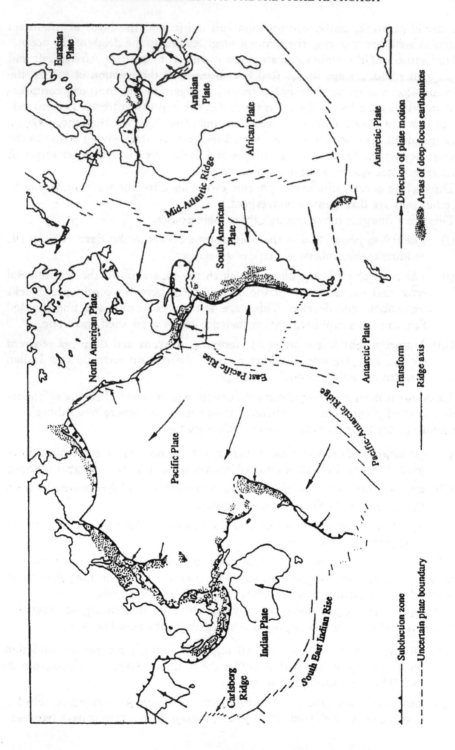

Figure 2.3 *The assessment of lithospheric plates*. From Kennett, 1982, 132.

(iii) Continental shelves cover about 7 per cent of the sea floor (27 million sq km). They are typically shallower than 130 m, but they can range up to 550 m.

(iv) Continental slopes overlay about 9 per cent of the ocean floor (28 million sq km) and range from the shelf edge to depths of 4000 to 5000 m.

(v) Continental rises, consisting of immense accumulations of terrigenous sediment, cover 19 million sq km of the ocean floor. They occur where continental margins lie within crustal plates so they do not occur in the convergent margins (Pacific type).

2.3 THE DYNAMICS OF THE WATER COLUMN IN SEA MANAGEMENT

Plate tectonic theory has become one of the most significant issues in research on sea management because of its capability of offering a comprehensive view, explaining *inter alia* the dynamics of the ocean bottom and the creation of mineral deposits. When the water column itself is investigated with the aim of providing frameworks for sea management a wide range of issues concerning both the physical and chemical properties of waters and their circulation emerge [Anderson A.T., 1982]. Three needs have acquired more and more importance.

Subject 2.2
The water column

The water mass, endowed with a salinity from 33 to 38 parts per thousand, could be thought of "as a gigantic pump that transfers heat from the equator to the poles. This transfer is affected in the surface waters of the ocean by strong currents, such as the Gulf Stream moving warm tropical waters to polar regions. The deep waters of the ocean have their origins in the high latitudes. Hence the deep waters are considerably colder than the surface water." [Turekian, 1976, 29]. As far as the use of ocean resources is concerned, it is worth recalling that the temperature of ocean waters, due to the compression of atmosphere and other local factors, changes vertically bringing about a transfer of heat from the surface to the deep layers. The temperature profile of the water column shows three distinct parts:

(i) *surface layer*, the temperature of which depend on the atmosphere compression on the water mass;

(ii) *layers from 100 to 1,500 m*, in which the temperature generally decreases;

(iii) deep layers and bottom influenced by the temperature of waters coming from high latitudes.

First, there is the need *to investigate physical and chemical properties on regional and local scales*, in order to provide knowledge necessary for the setting up of plans and management patterns. As the human impacts on the coastal zone have grown and diffused, the eutrophication process, as well processes caused by the variation of temperature altering food webs, have attracted research to the point that they have become the main concern in closed and semi-enclosed seas. At the present time the need to develop comparative research in intertropical, temperate and polar latitudes is regarded as crucial in order to sketch comprehensive theories about those processes. In this, the managerial world agrees with the scientific one on giving impetus to research on the change in physical properties and chemical composition of the water mass because these have strong implications for both resource and environmental management.

Secondly, the *need to have detailed analyses of circulation*. Bearing in mind that both horizontal movement and advection have a central role for fisheries and, *sensu lato*, for the exploitation of biological resources, sea management requires detailed knowledge of water mass dynamics on regional and local scales. At this level research provides increasingly numerous investigations and models of circulation in waters superadjacent the continental margins, particularly enclosed and semi-enclosed seas.

Subject 2.3
Circulation in semi-enclosed seas: the Western Mediterranean case

The complexity of the marine circulation in the Western Mediterranean Sea recently has been discussed by Millot through a model based on (i) the Modified Atlantic Water (MAW) on which the surface circulation depends; (ii) the Levantine Intermediate Water (LIW) forming the marine circulation in the intermediate layers; (iii) the Mediterranean Deep Water (MDW) bringing about the circulation in the deepest layers. Their features are represented in the Figure 2.3. The most important feature required to understand the water dynamics in this marine space—and in the Mediterranean as a whole—is seasonal variability of surface circulation, especially in the two areal currents: the Northern Current and the Algerian Current. [Millot, 1989].

Thirdly, the *need to explain changes in circulation*. As far as ocean circulation is regarded on the planetary scale, one of the most intriguing issues—strictly tied to sea management—is change in large-scale oceanic currents. The nature of this change, the potential relationships between it and change in atmospheric circulation and the subsequent effects on the availability of biological resources are investigated in order to relate these changes to surface and deep currents, ensuring that the main goals of resource management and environmental protection are appropriately pursued.

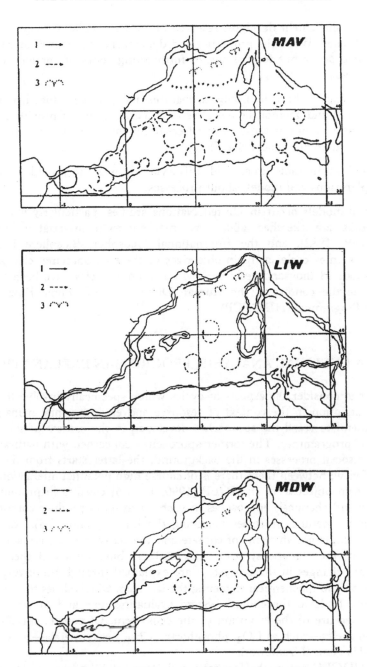

Figure 2.4 *Water circulation in the Mediterranean Sea*: MAW, Modified Atlantic Water; LIW, Levantine Intermediate Water; MDW, Mediterranean Deep Water. Key: 1. general permanent circulation; 2. permanent areal current; 3. winter areal current. From Millot, 1989, with adjustments.

As sea management develops, research on ocean circulation acquires increasing significance. When the properties and dynamics of the water column are not investigated *per se* but for the end of formulating codes of conduct and plans, analysis tends:

(i)   to consider both hydrological features and processes together with geological and geomorphological aspects with the aim of providing comprehensive views of those physical environments in which sea uses and environmental impacts take place;

(ii) and, at the same time, to identify cycles in order to provide dynamic explanations and models about processes.

Cyclical models occur in United Nations studies, particularly when general frameworks are sketched about the man-environment relationship at sea [GESAMP, 1982], and the International Geosphere-Biosphere Programme (IGBP) assumes cycles as main objectives of the environmental changes which began about 18 thousand years ago. More recently cycles have been taken into account in the context of the Human Dimensions of Global Environmental Change Programme (HDGECP).

2.4 CYCLICAL ANALYSIS: PROGRESS IN EXPLANATION

In order to consider the aspects of cycles which specifically involve the sea and call for attention in the context of resource and environmental management it seems useful to recall the approaches developed in the context of the IGB and HDGEC programmes. The former approach is concerned with natural sciences and puts social processes in the background; the latter starts from the opposite point of view; between them there is of course high potential interaction.

The main objective of the IGBP [1990, 1.3-1.6] consists in promoting joint (physical and chemical) approaches in such a way as to produce comprehensive views of the natural processes involving the Earth (Figure 2.5). Natural processes are explained in terms of climate and the ecosystem. As a result, analysis starts from the climatic system, deals with its biophysics and then with the ecosystem; changes in the ecosystem are explained through the involvement of biophysical and biochemical elements; the last step, which closes this circular relationship, leads to the climate system evaluating how and to what extent it changes because of the evolution of the ecosystem. Of course, specific biogeochemical cycles—such as CO_2, phosphorus, CH_4, heavy metals—are considered within this general framework.

The HDGEC approach [Jacobson and Price, 1990] allows us to investigate the inputs moving from social processes and leading to the climate system and the ecosystem through the consideration of (i) the disturbances in natural systems and (ii) those social processes which behave as agents "of biospheric and

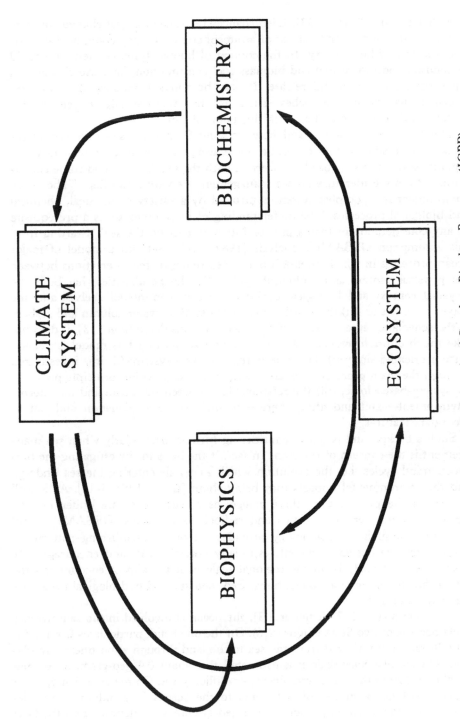

Figure 2.5 *The general approach of the International Geosphere-Biosphere Programme* (IGBP).

geospheric change" [*Ibid.*, 33]. Disturbances relevant to global change concern, of course, the climate and consist of stratospheric ozone depletion, acid deposition and loss of biodiversity. In the context of human systems attention should be focused upon fossil fuel and biomass fuel consumption, land use change and halocarbon production and release. It may be noted in passing that some regional management approaches—such as that for the Mediterranean Sea [UNEP, 1989]—are consistent with this point of view.

The question is whether and how, moving from these scientific considerations, it is possible to consider cycles generating changes on both planetary and regional scales and specifically concerned with the sea. To this end some contributions from the literature on sea management are worth recalling. "The ocean environment is a complex system controlled by a variety of physical, chemical and biological processes. The understanding of these processes is a prerequisite of any consideration of man's past or future impact on the sea". Starting from this assumption GESAMP's analysis [1982, 11-13] sets up a model of major ocean processes in which attention is focused upon: (i) the interactions between the physical context and biological cycles; (ii) the peculiarities that both the physical context and biological cycles acquire in continental shelves and the "open ocean", and (iii) the peculiarities of sea surface, water column and seabed. If the schematic representation of this model is considered [*ibid.*, 13], it does not take much effort to realize that the paradigm supporting this model is general system-oriented since: (i) the ocean is regarded as a system; (ii) it is investigated through the main processes, i.e. according to the aggregation principle provided by system-based logic; (iii) the relationships between the ocean and its external environment—land and atmosphere—are included in explanation and not as secondary features.

Such a background is worth considering further, particularly when sea management is the concern of research. In fact, it the basis for investigating the biogeochemical cycles, i.e. the evolution which materials entering the sea undergo and the environmental impacts they bring about [Clark, 1986, 1-10]. As is well known, at the present time three categories of substances are under investigation: nutrient elements, metals and radioactive materials. GESAMP [1982] suggests that such a methodology provides a tool for establishing patterns to manage the marine environment, especially coastal areas, and encourages the development of analysis on the regional scale with the aim of encouraging the establishment of regional conventions. Sedimentary and hydrological cycles are the main concerns.

As Odum states [1971, chapter 4.3], the ocean is involved in the sedimentary cycle because of two flows (Figure 2.6). On the one hand, particulates from natural fall-out are transferred from the sea to the land through atmospheric circulation. On the planetary scale it is estimated that about 0.4 geo-grammes of material moves from the sea to *terra firma* per million years. On the other hand, rivers transfer 184 geo-grammes of sediments to the oceans in a million years, deposited on the seabed: in particular, 123 geo-grammes are deposited on the floor

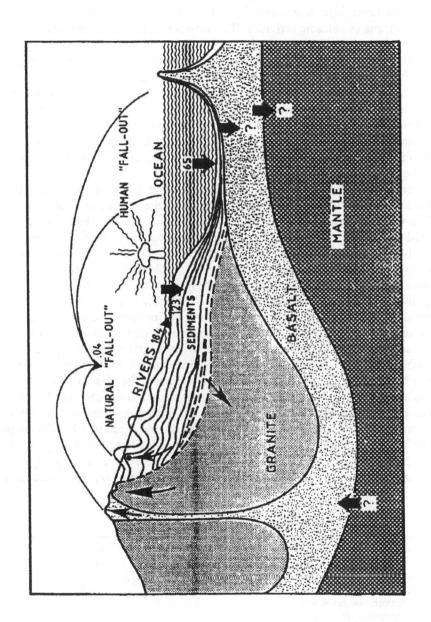

Figure 2.6 *The sedimentary cycle.* Flows are expressed in geogrammes per million years (1 geogramme = 10^{20} g). From Odum, 1971.

of the coastal sea—so they involve the continental margin—and 65 geo-grammes reach the deep seabed. It is less well known how materials move from the sea bottom to the basalt layer below and from which they are transferred to the land surface by means of volcanic activities. The interaction between the basalt layers and the mantle are less well known.

The relationships between the sedimentary cycle and plate tectonic dynamics—which, in its turn, is based on the continental rifting, drifting and reaggregation [Chappell, 1987, 50]—are self-evident. As far as sea management is concerned, research gains interest when is developed on the regional scale with the aim of knowing how and to what extent the transfer of sediments from land to the sea, and *vice versa*, is influenced by the dynamics of plates and the peculiar features of margins. The need to shift from the planetary scale to the regional one is one of the most methodologically exacting tasks.

The complexity of regional sedimentary cycles and other factors generating changes in coastlines [Bird, 1985] is growing since the human pressure on the shoreline and continental margins grows and diffuses and the river basins are more and more affected by human settlements and activities. In fact, as the actions of both coastal and inland factors become greater, the sedimentary cycle undergoes alterations. This leads to a range of problems on the regional scale: the coastal areas that, at the present time, are largely influenced by this process are significant case studies. As a first approach, attention is attracted by littoral megalopolises [Stewart, 1970], in which urban and industrial settlements cover such large areas as to strongly influence geomorphological dynamics. Both the Atlantic and Pacific megalopolises in the United States, those on the eastern side of Honshu and those adjacent to the North Sea and the northern side of the Western Mediterranean Sea, may be regarded as interesting case studies.

When attention moves to the dynamics of water masses the hydrological cycle is the main issue. The literature has provided models of varying degrees of complexity concerning this cycle, which is acquiring a crucial role in relationships between man and the environment. The model considered in *The State of the Planet* [King, 1980, 45], the well known report of IFIAS, can be regarded as appropriate for sea management-oriented analysis (Figure 2.7). According to this view, the ocean as a

TABLE 2.2

Water balance on the Earth's surface

processes	grams per year $\times 10^{20}$
evaporation from ocean	3.83
precipitation on ocean	3.47
evaporation from land	0.63
precipitation on land	0.99
runoff from land to sea	0.36

From Turekian, 1976, 11.

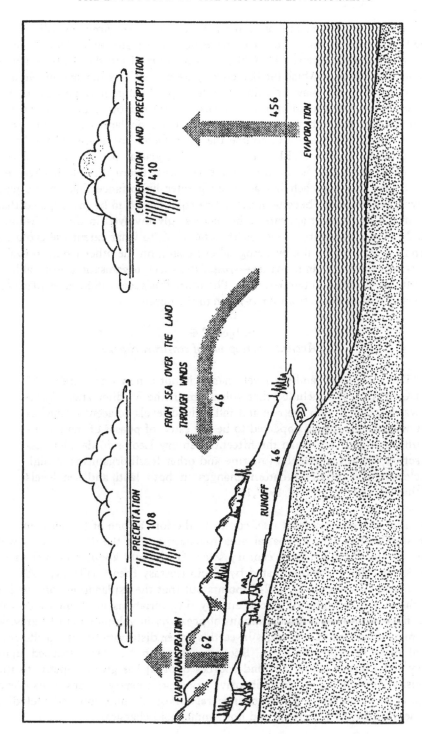

Figure 2.7 *The hydrological cycle,* expressed in 1000 cu km/year. From M.I. Budkyo, 1974, reprinted in King, 1980, 45.

whole receives 46,000 cu km of water a year from *terra-firma* by runoff and 410,000 cu km from the atmosphere by condensation and precipitation. In its turn, the atmosphere receives 456,000 cu km of waters from the sea by evaporation, about one tenth of which (46,000 cu km) are carried over the land by winds.

The ocean does not play a significant role in producing alterations in the hydrological cycle. On the contrary, it is largely affected by changes which this cycle undergoes through the processes taking place on land and in the atmosphere. Climatic changes are of course the main factor considered in the literature. According to the IGBP, the current post-glacial period, which started about 18,000 years ago, is the time reference basis. From the sea management point of view this approach has acquired greater importance than in the past. However, this cycle has become much more complicated to be investigated than it was in the 1970s, as it appears to be increasingly involved in the green-house effect. Now it is necessary to see, on the one hand, to what extent the evolution of climate is influencing the hydrological cycle and, on the other hand, to evaluate to what extent human activities—*sensu lato*, social organisations—are able to accelerate the increase in temperature. The more this acceleration takes place the more water masses move from *terra-firma* to the ocean.

Subject 2.4
The Ancient's perception of sea-level change

"The concept of land/sea-level change is not a new one. Strabo (68 BC to AD 24) concluded that volcanism in the Mediterranean basin was associated with the rise and fall of land levels adjacent to the sea and that this process appeared to be the cause of phases of marine inundation and retreat. In the fifteenth century Leonardo da Vinci asserted that the presence of marine and other fossil organisms at high elevations is proof of datum changes in both land and sea levels through time". [Devoy, 1987,1].

A crucial concern, sea-level rise, comes to the fore as one of the most important research objectives in coastal area management in the 1990s. As is well known, during the nineteenth century scientific thought about sea-level variation [Tooley, 1987] was dominated by glacio-isostasy theory [Devoy, 1987, 5]. In particular, Mörner [1987, 332-7] points out that the explanation consisted of three theories concerning (i) glacial eustasy (i.e. variations in ocean water volume), (ii) tectono-eustasy (i.e. variations of ocean basin volume) and (iii) gravitational mass attraction (i.e. changes in earth's water distribution) due to the continental mass and the ice mass. Recently attention has been focused upon isostasy and Mörner [1987, 338 and sources quoted] has given impetus to this propensity introducing a comprehensive concept: isostasy is any ocean-level change or any absolute sea-level change regardless of causation and including both the main family of vertical and horizontal geoid changes.

This concept leads us to take into account four categories of factors influencing sea-level variations (Figure 2.8):

(i) *tectono-eustatic variables*, which influence the ocean basin volume: both orogenetic factors, mid-oceanic ridge growth, plate tectonics, etc., and local and hydro isostasy, etc., are the main issues;

(ii) *glacial eustatic variables*, which influence the ocean water volume;

(iii) *geoidal eustasy*, which depends on gravitational waves, tilting of the earth, earth's rates of rotation and deformation of the geoid and brings about changes in the distribution of both ocean waters and sea-level;

(iv) *meteorological and hydrological factors*, which converge to produce sea-level changes, especially on the regional and local scales.

EUSTASY OCEAN LEVEL CHANGES	VERTICAL AND HORIZONTAL GEOID CHANGES			EARTH - VOLUME CHANGES	
		OCEAN BASIN VOLUME	TECTONICS	OROGENY	
				MID - OCEANIC RIDGE GROWTH	
				PLATE TECTONICS	
				SEA FLOOR SUBSIDENCE	
				OTHER EARTH MOVEMENTS	
			ISOSTASY	SEDIMENTS IN - FILL	
				LOCAL ISOSTASY	
				HYDRO - ISOSTASY	
				INTERNAL LOADING ADJUSTMENT	
		OCEAN WATER VOLUME		GLACIAL EUSTASY	
				WATER IN SEDIMENT, LAKES AND CLOUDS, EVAPORATION, JUVENILE WATER	
		OCEAN MASS / LEVEL DISTRIBUTION	GEOIDAL EUSTASY	GRAVITATIONAL WAVES	
				TILTING OF THE EARTH	
				EARTH'S RATE OF ROTATION	
				DEFORMATION OF GEOID RELIEF (DIFFERENT HARMONICS)	
	DYNAMIC CHANGES	DYNAMIC SEA LEVEL CHANGES		METEOROLOGICAL	
				HYDROLOGICAL	
				OCEANOGRAPHIC	

Figure 2.8 *The eustatic variables*, according to Mörner (1987, 339).

This approach appears stimulating also when analysis is focused on the inter-action between physical and human factors, as well as oriented to provide expla-nations for semi-enclosed seas [Pirazzoli, 1987]. One wonders whether scientific thought may provide models which, starting from Mörner's approach, are able to deal with impulses arising from human settlements and activities and, *sensu lato*, from planning and management. In any event, when coastal management is considered, the timescale should also be introduced: some of factors produce their influences in the long term, e.g. on plate tectonic time scale, other factors act in the medium term, e.g. on the recent climatic change time scale, and other factors act in the short term, namely, on an historical time scale. In this context attention shifts from the list of the Mörner model's factors to a process that in-volves both the volumes of the ocean basins and the ocean water features. In this context two subjects gain interest: (i) the role of endogenous and exogenous fac-tors producing changes in the sea mass and (ii) the occurrences of real and ap-parent changes [Devoy 1987, 14-15].

2.5 THE RESPONSES FROM THE EVOLVING MARINE ECOSYSTEM

Geological and geomorphological processes, on the one hand, and chemical pro-cesses and cycles, on the other, interact with the marine ecosystem. As already noted in Chapter 1, the general idea of the ecosystem consists of a system of liv-ing elements that interacts with a abiotic set of elements. This idea can be ac-cepted when attention to ecosystems is due to the need to build management patterns. In order to distinguish (i) biological elements and food chains from (ii) their abiotic bases it will afterwards be of the *marine ecosystem* and its *physical environment*. During the past, particularly before the 1980s, marine ecosystems were investigated and classified mainly in terms of food chains and food webs. More recently research has been focused on (i) the structure of the ecosystem and (ii) the interaction between the ecosystem, on the one hand, and its physical context and the external environment, on the other. "After about a decade and a half of experience with multi-level process models, there is a fairly widespread view that the attempt to reproduce the behaviour of the whole by putting to-gether the properties of the parts may never succeed. The natural system has self-regulating (homoeostatic) integration and behaviour that result from feed-back, so that to model all of the processes makes the model too complex to manipulate" [IOC, 1984, 65].

Both in the analysis of terrestrial [IGBP, 1990, Chapter 6] and marine ecosys-tems [IOC, 1984, 64-6] efforts have been made to understand their physiology. The idea of physiology presupposes that of pathology of the ecosystem, which has been investigated distinguishing natural causes [Clark, 1986, 6-10] from those due to human behaviour [IOC, 1984, 67-8]. The implications of marine ecosystems in human behaviour have been explained in terms of inputs from man to the sea and of responses of the ecosystem. These responses display two aspects: endogenous (the ecosystem) and exogenous (its external environment). *Responses in the endogenous context* bring about changes in the structure of the ecosystem leading to simple adjustments or profound alterations, which—according to general system theory—can be described in terms of morphogeneses. *Responses in the exogenous context* involve the natural environment (e.g., physical and biological processes) and direct impulses of human communities (e.g., altering the possibility of exploiting physical resources, jeopardizing health).

Responses and alterations depend, *inter alia*, on the kind of ecosystems and their dimensions. To cluster the range of possible responses from the ecosystem to inputs in which it undergoes, a more or less large number of variables can be taken into account but it cannot leave out two of them: geomorphological fea-

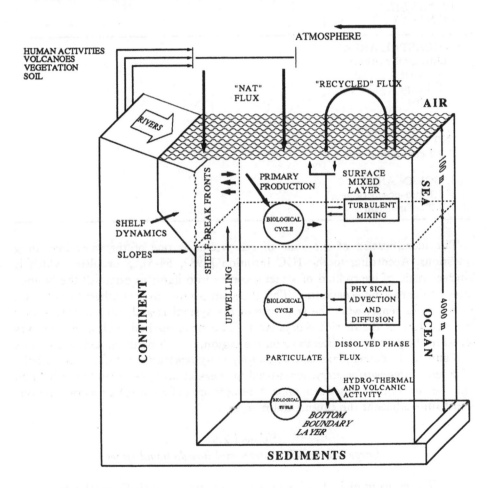

Figure 2.9 *The model of major natural marine processes.* From GESAMP, 1982.

tures and latitude. The former variable (geomorphology) leads to distinction between the ecosystems of the continental margins and the ocean areas. The continental margins can be disaggregated into the shelf, slope and rise, and the ocean areas into mid-ocean ridges, ocean basins and continental rises (regarded as interfaces between the margin and the ocean). The latter variable (latitude) suggests distinguishing the ecosystems of the intertropical basins from those of temperate belts and those of sub-polar and polar areas.

As a result, this first approach framework comes to the fore.

TABLE 2.3
Types of marine ecosystems

LATITUDES: PHYSICAL CONTEXT	intertropical	temperate	sub-polar	polar
1. COASTAL AREA (continental margin) 1.1 shelf 1.2 slope 1.3 rise				
2. COAST/OCEAN INTERFACE 2.1 continental rise				
3. OCEAN AREA 3.1 mid-ocean ridge 3.2 ocean-basin				

The identification of the scale of ecosystems is one of the most intriguing questions. According to the IOC intuition [1982, 64-5], a problem, which is characteristic of every kind of system, comes into light: to establish the boundaries of the marine ecosystems. This problem presupposes another: to establish, in a general sense, what the structure of the system is and, subsequently, what the structure of the marine ecosystem is. The same methodological concern was faced in the past by the literature on the region, as will be discussed in Chapters 6, 8 and 10. In current language *the marine ecosystem* means ocean life as a whole. The term *large marine ecosystem* is used to speak of ecosystems on the scale of an enclosed or semi-enclosed sea, an archipelagic area of a part of a marine environment superadjacent the continental margin.

Subject 2.5
Large Marine Ecosystems and jurisdictional issues

"The concept of LMEs involves several offshore jurisdictional zones. The landward side is the territorial sea (and in some cases also internal waters) where, for resource purposes, the coastal state has complete sovereignty. Beyond territorial limits, most coastal states claim and Exclusive Economic Zone (EEZ) or Exclusive Fishery Zone. Here again, for resource purposes, the coastal state has sovereignty, subject, according to the 1982 Law of the Sea (LOS) Convention, to certain rights by foreign states, such as the right of access to 'surplus' fisheries. Seaward of the EEZ are the high seas where all states have the right to participate in the fisheries. Some LMEs do not meaningfully encompass areas of the high seas" [Alexander, 1989, 340].

Recently researchers have tried to define the Large Marine Ecosystem and to formulate a theory about it. According to the approach that has supported the Conference on "The Large Marine Ecosystem (LME). Concept and its Application to Regional Marine Resource Management" (Monaco, 1990), "Large marine ecosystems are defined as large regions of the world ocean, generally around 200,000 square kilometers, characterized by unique bathymetry, oceanography, and productivity within which marine populations have adapted reproductive, growth, and feeding strategies, and which are subject to dominant forcing functions such as pollution, human predation and oceanographic conditions. Many of these LMEs worldwide are being subject to stress from various forms of human interference or uses, including pollution and heavy exploitation of renewable and non-renewable resources against a background of growing global change". This concept is the current product of a theoretical evolution, well supported by analysis of case studies, which began in 1984 and has recently reached interesting approach [Alexander, 1986 and 1989]. It is policy-oriented because its main goal is to provide a concept able to justify plans for resource and environmental management of (i) the exclusive economic zone and (ii) enclosed and semi-enclosed seas or their large parts. At the present time it is difficult to predict what scientific results will be achieved through this approach, which—as can be seen—should deal with relevant theoretical and methodological problems. Two deductions seem to be justified. First, this approach is typically goal-oriented. Secondly, it has explicitly introduced the ecosystem as an unavoidable aspect of marine regionalization. Prospects, both in ecology- [Morgan, 1989/b and 1990] and in policy-oriented [Prescott, 1990] researches are meaningful.

CHAPTER 3

THE EVOLUTION OF SOCIETY

3.1 THE EVOLUTION OF SOCIETY: HISTORICAL STAGES

On the historical timescale, the interaction between social and natural processes is the focus of interest (Figure 3.1). Natural processes can be explained through cycle-based models, so the basic idea is that of the ecosystem which evolves passing through bifurcation phases and undergoing radical adjustments or morphogeneses. Social processes are currently described through stage-based models in which social organization and management of the earth's resources move from one stage to another: the passage between stages is the result of the choice that societies accomplish within a spectrum of opportunities. In the short term, the transition to a new economic and social stage occurs in a bifurcation phase. As has been stated (Chapter 1), the idea of bifurcation can be assured as an isomorphism through which both natural and social processes can be described [Laszlo, 1988].

In this view reasoning may proceed through four steps: (i) the historical stages, which provide a basis for taking into account the evolution of specific sea use categories; (ii) littoral industrialization, regarded as one of the main processes that, in recent times, have involved coastal areas; (iii) the evolution of some sea uses that have played a special role in the present historical phase (as an example, navigation and the offshore oil and gas industry).

As far as historical stages in sea management are concerned, attention may be centred on those which took shape from the pre-industrial phase, namely, from the enlargement of the European world view due to the discovery of America and afterwards (15th-18th centuries). In this context Geddes [1915] and Mumford [1934] constructed historical models having two important features: (i) great importance was attributed to the interactions between technological advance, social organization and spatial management; (ii) society as a whole was concerned so the content of models was comprehensive. Geddes' approach is focused upon two stages: palaeo-technical and neo-technical. Mumford's model leads to the identification of three stages: eo-technical, palaeo-technical and neo-technical. More recently, Rostow [1960] constructed a stage-based model specifically devoted to the explanation of economic and social processes involving

46

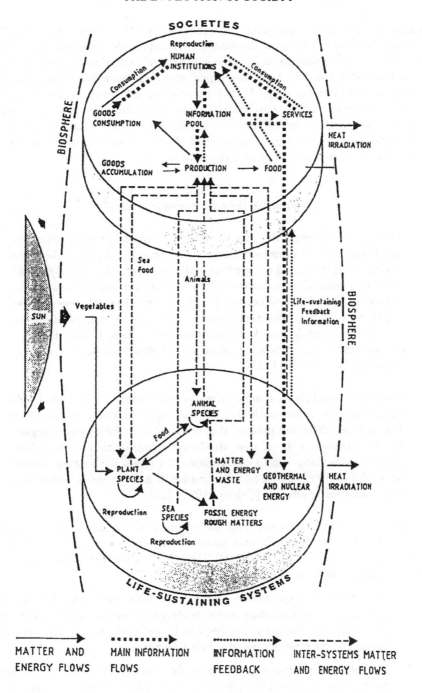

Figure 3.1 The interaction between society and the natural environment. From Laszlo, 1987, 386.

nations. In particular, analysing growth rates and the resource accumulation processes leading to industrial development, Rostow identified five stages: traditional society and preparation for take off, take off, drive to maturity, maturity. According to this model, well developed societies move from maturity to the high mass consumption stage. It may be stressed that Geddes' and Mumford's models, on the one hand, and Rostow's model, on the other, are supported by very different methodological approaches [Keeble, 1967]. In addition, it is clear that, as far as global change is concerned, Rostow's model, formulated on the national scale, is much less relevant than the others.

Subject 3.1

Changes in the ecosystem and in society

"Change in ecosystems are twofold: (a) rapid changes (disasters, catastrophes) that destroy structures and that are unpredictable from inside the system, and (b) gradual and slow changes that vanish gradually in a generalized situation of low energy and that admit a considerable degree of local diversification (...). The change is towards segregation, and the persistent situations consist of misplaced things (...). In the oceanic world of plankton, where there is light there are no nutrients: nutrients accumulate in the depths, where there is no light." [Margalef, 1985, 234].

"Organizational social change refers to the change (...) in the (..) rules of conduct. If, under such change, the individuals remain capable of maintaining their autopoiesis, they will recreate the changed organization together with its new structural manifestations—the change will be viable, long-lasting, and capable of its own self-renewal. Otherwise, the change may lead to the demise of a given social system as a self-maintaining entity". [Zeleny, 1985, 323].

Starting from models concerning society as a whole and focusing attention upon ocean involvement in human uses and activities, a stage-based historical model is proposed [Vallega 1990/a, 69-72], which is based on four stages: mercantile, palaeo-, neo- and post-industrial (Figure 3.2).

(i)   The *mercantile stage*, developed from the 16th century to the end of the 18th century. In this stage world trade grew because of the ocean routes opened up by the great geographical explorations and the subsequent changes in the international division of labour.

(ii) The *palaeo-industrial stage* was brought into existence by the first industrial revolution, took off during the second half of the 18th century and entered its maturity phase in the second half of the following century. Thanks to the discovery of techniques to convert thermal to mechanical

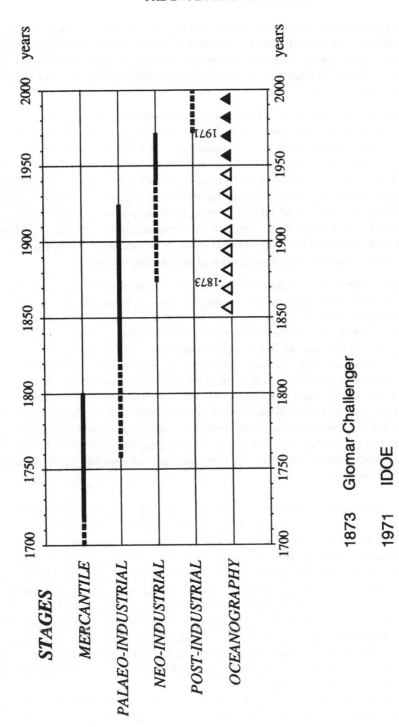

Figure 3.2 *The stage-based historical model* from the phase which led from the palaeo-industrial stage to the post-industrial one. From Vallega, 1990a, 70.

energy, profound changes in the division of labour in manufacturing activities were produced and the international division of labour was transformed to the point of bringing about revolutionary changes in maritime trade. The world began to be divided into industrialized and non-industrialised countries

(iii) Based on the techniques of converting thermal into electric energy, on the exploitation of oil and gas and, more recently, on the use of nuclear fuels, the *neo-industrial stage* emerged in the United States in the late 19th century and from the 1920s spread out to other parts of the world—such as in Western Europe and Japan. As had happened in the previous stage, because of a wide range of technological inputs, changes first took place in the division of labour of manufacturing activities and in social contexts. Afterwards they involved the international scene, giving birth to a new pattern of the international division of labour and brought about the dichotomy between developed and developing countries.

(iv) It is a matter for discussion when and how the *post-industrial stage* came to the fore in ceasing some important growth processes sustained by mechanisms generated by the neo-industrial economy. However, it is agreed that the first wave of inputs developed in the late 1960s and in the early 1970s. Changes in the international division of labour—in their turn produced by relevant political processes, such as de-colonisation—had a much more important role than they had had in the previous stage [Post, 1983, 66-68]. As far as technological processes are concerned, changes in energy systems have still not played an important role, while information systems and related telecommunication advances are contributing much more than could have been foreseen in changing the division of labour in firms and societies, as well in the international context [Interfutures, 1979, 114-50].

This stage-based historical model can be regarded as a preliminary background for the identification and explanation of bifurcations in social activities at sea—and, *sensu lato*, of the role of the sea in human development. During the sequence of historical stages some changes have occurred in the marine context.

The sea in the mercantile stage. Before the rise of the palaeo-industrial stage the sea was affected by a limited range of uses: rowing and sailing, navigation, fishing, extraction of salt and military activities on the sea surface. In truth, until that stage the exploitation of marine resources—and human activities on the sea—had changed very quickly: century by century they underwent only adjustments. The main advances had taken place in navigation, which had benefited from some new technical tools. In particular, the human presence on the sea had not affected marine or coastal ecosystems.

The sea in the palaeo-industrial stage. As a main consequence, the bifurcation which led to the palaeo-industrial economy brought about the development of the coal-fired vessel, which gave birth to the initial consistent impact on the marine environment. Submarine telegraph cables were introduced, which was the first impact on the seabed caused by the technological advance in communication. Extraction of gravel and sand diffused because of the need for materials to build seaports and coastal settlements. Lastly, research had its debut including the employment of oceanographic vessels and instruments to measure water depths, to explore the seabed, and to investigate physical properties of water masses. As can be seen, both in ocean and coastal areas the human presence changed and new goals in resource exploitation were pursued.

The sea in the neo-industrial stage. Much more impressive was the bifurcation that led to the neo-industrial organization. The oil-fired vessel was introduced; an important network of telephone-cables was placed on ocean bottoms; industrial fishing and aquaculture came to the fore; and submarine military activities spread out. The take off of the neo-industrial one was marked by two features *vis-à-vis* of previous historical stages. First, there was profound technological advances in sea uses: e.g. the organization of manufacturing processes based on refrigeration radically changed fisheries. Secondly, the water column was involved in new kinds of uses, such as submarine navigation, offshore oil and gas installations, recreational activities, and salt extraction. Because of this new spectrum of techniques and management patterns, such as military activities, the sea was exposed to hazardous uses, particularly for ecosystems.

The sea in the post-industrial stage. During the 1960s, while the neo-industrial economy was also spreading out in many developing countries, the sea was affected by new impulses (Figure 3.3). These were so many and so deeply influenced the marine environment that the sea entered a bifurcation phase, which has developed during the 1970s: information systems and robotics-equipped vessels have become a revolutionary means of navigation; oil and gas offshore exploration and exploitation have been the most impressive field of technological advance in marine engineering; coastal marine areas, such as Japanese bays, have been used for settlements; submarine archaeology was born and has been practised in ever marine areas; submarine parks have opened the way to cultural sea uses, such as submarine tourism; also thanks to robotics, research has developed in deeper and deeper waters. Technologies have been provided for a significant number of other uses including the production of energy from the thermal gradient in the water column; the exploitation of deposits of manganese nodules; the extraction of salts from phosphorite rocks; and settlements on the sea bed. Lastly, the perception of the ecological implications for the ocean has stimulated a new range of technologies and activities aiming at conserving and preserving marine ecosystems and their physical contexts. In conclusion, these occurrences demonstrate that (i) because of the transition from the neo- to the post-industrial stage, the sea has been profoundly affected by human presence and has been involved in increasing exploitation of marine resources; (ii) the technological and organisational advances in sea uses have a central role for the evolution of the post-industrial stage.

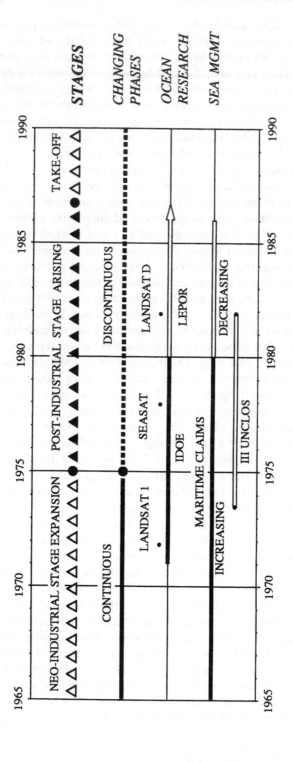

Figure 3.3 *The transition phase from the neo-industrial to the post-industrial stage, and some important events in ocean research and management.* From Vallega, 1990a, 74.

In particular, the present bifurcation phase is marked by at least three features:

(i) because of the extension of uses and their increasing importance for mankind, the sea is acquiring comprehensive management patterns;

(ii) while in the past most technologies were transferred from land activities to sea uses, at the present time the contrary occurs (e.g., the technological advance in oil and gas offshore exploitation is useful also for land exploitation;

(iii) what takes place at sea influences not only marine ecosystems but also land and atmospheric environments, particularly because of the hydrological cycle.

The features of this transition phase lead to consideration of natural and social processes together, with the aim of understanding their growing interaction due to the diffusion of the human presence at sea and the implementation of uses. In this context three timescales are to be considered:

(i) the *plate tectonic-based scale*, because it is necessary to put an important range of implications—such as volcanism and seismicity—in relation to the dynamics of convergent, divergent and subducting plates;

(ii) the *climatic phase—based scale*, by which attention is focused on what has occurred from the maximum extension of the WÅrm glaciation, i.e. about 18,000-20,000 years ago, until the present time;

(iii) the *historical scale*, which is centred on the stages following the first industrial revolution.

3.2 THE SPATIAL DIMENSION: DIFFUSION PROCESSES

The spatial diffusion of such things as technologies and organisational patterns, resource exploitation models, and environmental implications, is closely tied to the bifurcations occurring in economic and social contexts. This is not only true *sensu lato* but also *sensu stricto*, i.e. in relation to the involvement of the sea in economic and social processes. As far as this particular context is concerned, diffusion processes which emerge include:

(i) *technologies*, because changes in sea uses and in environmental management are both closely dependent on technological advance and diffusion;

(ii) *uses*, bearing in mind that any bifurcation brings about new ways of managing conventional uses and creates new uses;

(iii) *perception*, because the evolution of sea uses and subsequent environmental implications allow human communities to perceive the role of the sea in a different way from the past.

It may be emphasised that these processes are reciprocally linked by feedback relations:

(i)   the more technologies advance, the more sea uses grow and diffuse and, in their turn, give impetus to technological advance;

(ii) the more new kinds of sea uses come to light the more the perception of the role of the sea in the destiny of mankind develops and, in its turn, encourages the implementation of sea uses.

Literature provides a framework for diffusion processes, illustrated by Hägerstrand's diffusion model [1953]. According to this, the diffusion of innovation passes through four phases: (i) the *take off phase*, in which innovation develops in some places, generating spatial differentiation between these and the other spaces; (ii) the *diffusion phase*, in which innovation diffuses in a centrifugal way so that an expanding area is involved and spatial differentiation decreases; (iii) the *consolidation phase*, in which ever more extended areas adopt the same resource management patterns and technologies; (iv) the *saturation phase*, in which this management and technological endowment reaches first its maturity and then its senility.

Models based on this framework can be taken into consideration but, in the meanwhile, any diffusion-oriented model should be evaluated in relation to three concerns: the *time scale*, appropriate for dealing with natural and social processes together; the *spatial scale*, by which attention is focused on the methodological implications relating to the macro-, meso- and micro-areas; the *bifurcation phases*, which should be taken into consideration with the specific aim of knowing which processes lead to morphogenesis both in ecosystems and sea management patterns. This preliminary reasoning scheme—based on the thought of Braudel [1979]—can be considered.

When a new stage is about to start a bifurcation phase takes shape in a given area—e.g., Great Britain for the palaeo-industrial stage, the United States for the neo- and post-industrial stages. This is the *generating area*.

Impulses move from this area to other countries and regions which, in their turn, acquire importance because of the spatial differentiation produced by innovation waves. The set of countries and regions involved by this process makes up the *leading area*.

The rest of the world consists of (i) countries and regions which are not affected by this diffusing change and (ii) countries and regions which are asked to provide resources—raw materials, energy sources, manpower, etc.—to the leading area.

In this context both the generating area and the leading area converge to produce a new international and inter-regional division of labour.

Inter alia, the transition from one historical stage to the next generates three important effects: (i) a more extended part of the world is involved in international relationships; (ii) new spaces are used for settlements and activities;

(iii) national systems are involved in a wider range of international relationships.

Since the first Industrial Revolution, the sea has not only been increasingly involved thanks to bifurcations giving birth to more and more numerous sea uses, but also the geographical focus of the ocean world has changed. During the palaeo- and neo-industrial stages the Atlantic Ocean, particularly its northern section, was the focus of the international economy—during the former stage because of the innovations diffused from Great Britain and during the latter because of the innovations diffused from the United States. From the 1960s, i.e. from the birth of the present bifurcation, an impressive development is taking shape in the Pacific Ocean, in which the western side of the United States, Japan and Australia are playing a leading role. As a result, the Pacific is the core of the present changes in the marine world.

3.3 COASTAL INDUSTRIALIZATION: GENERATIONS OF MIDAs

Moving from general views of the involvement of the sea in the evolution of economic and social characteristics to specific processes, coastal industrialization is worth considering. In the recent past the literature has made efforts to investigate this leading phenomenon and, particularly, to explain it by means of stage-based models.

Attention can be restricted to cognitive and predictive models of the industrial development in coastal areas. According to Vigarié [1981], the growth of the international economy has given rise to four generations of Maritime Industrial Development Areas (MIDAs; Figure 3.4).

The first generation, which developed in the Western World from the 1950s to the early 1970s, was based on a range of leading sectors, including the processing of raw materials and energy sources: the iron and steel industry, oil refining, production of aluminium, and thermoelectric plants. The coastal area as a whole was central to these developments, due above all to the large scale of the industrial processes involved. Scale economies had great importance because: (i) the need to lower manufacturing costs led to the creation of large MIDAs—some of them extended to more than 10,000 hectares; (ii) the demand for large and ultra large bulk carriers increased to the point of building 150,000 dwt solid bulk carriers and over 500,000 dwt tankers [Vallega, 1980, 112-118]; (iii) the increase of average deadweight tonnage of bulk carriers encouraged the enlargement of the MIDAs with the aim of taking the maximum advantage from the scale economies of the entire production cycle.

Coastal areas were involved in this process not only because of the growing demand for land surfaces and other physical resources—such as water, sand and gravel, etc.— but also because of the need for large-scale waste disposal, which has implications for coastal ecosystems, shoreline and seabed morphology, and the physical and chemical properties of water. As far as offshore areas are concerned, the main environmental implications were due to the so-called

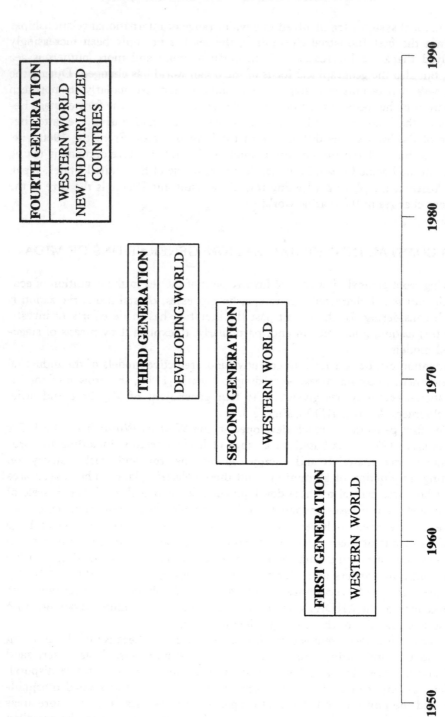

Figure 3.4 *The generations of Maritime Industrial Development Areas (MIDAs), according to Vigarie's model (1981).*

"giantism" of vessels, i.e. by the aforementioned growing deadweight tonnage of bulk carriers. The risk of pollution because of accidents and collisions increased in several parts of the oceans, particularly along sea routes from the Persian Gulf to the most economically developed countries of the Western World.

Vigarié [1981] stressed that in some regions, specifically in the most advanced ones—e.g. in the Netherlands—in the late 1960s and in the early 1970s a tendency to adjust this industrialisation pattern emerged. Strategies were developed to limit the expansion of primary production and to promote the creation of the manufacture of finished goods so as to reduce pollution and to increase the productivity of the exploitation of coastal resources. So a second generation of MIDAs, more careful in preventing environmental impacts, took shape, but in truth this occurred only in a few areas, such as in the Netherlands.

This adjustment process was diffusing when, in 1973-1974, the Organization of Petroleum Exporting Countries (OPEC) increased oil and gas prices stopping the growth of important manufacturing sectors—particularly the iron and steel industry and oil refining—in many parts of the Western World. In the meantime MIDAs started diffusing in developing countries with the objective of processing local minerals and producing semi-finished goods for the international market. In other words, a new MIDA generation, the third one, rapidly took shape, acquiring a functional profile similar to that of the first generation because it was based mainly on primary production.

Recently, in the late 1970s and 1980s, another significant process has involved coastal industrialization: in developed countries a growing number of conventional manufacturing activities have been transformed and other functions have been set up: both high technology-based production and tertiary activities are acquiring a leading role. In addition, as the exploitation of marine resources advances, these areas are endowed with facilities concerned with marine activities, such as offshore installations, aquaculture, and recreational uses. According to Vigarié's model [1981], this new pattern may be regarded as the fourth MIDAs generation. At the present time it is difficult to understand to what extent this model will advance and will interact with the implementation of sea uses. *Optimo iure* it is difficult to predict how it will be concerned with sea management. In any case, it may be considered as an important issue.

3.4 ORGANISATIONAL PHASES IN MARITIME TRANSPORTATION

While the transition phase from the neo- to the post-industrial stage was bringing about the decline of conventional MIDAs in the Western World and the diffusion of MIDAs in developing countries, some significant changes have occurred in mercantile navigation and transportation. First of all they took place in oil transportation.

(i)   In the mid 1970s the propensity to build larger and larger oil carriers suddenly stopped. The last product of this giantism in transportation

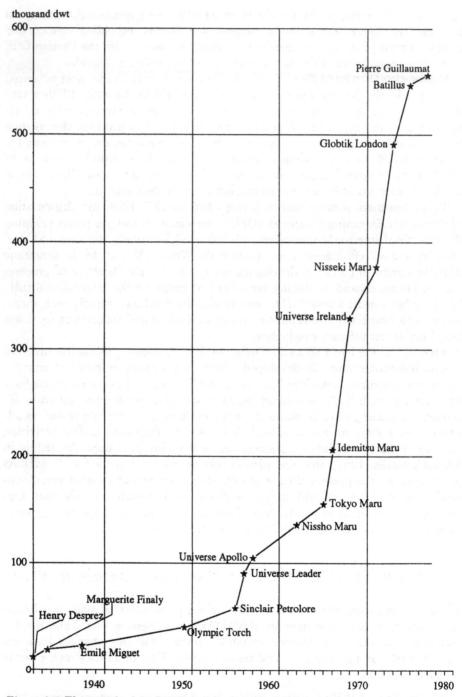

Figure 3.5 *The growth of deadweight tonnage in oil transportation* during the neo-industrial stage. From Vallega, 1990a, 103.

reached its peak in 1976, when the 540,000 dwt *Batillus* was launched. Afterwards the employment of very large and ultra large vessels has decreased, while the use of medium-sized carriers—i.e. less than 150,000 dwt—has increased. As a result, maritime oil transportation has lost one of its most hazardous features for both coastal and ocean environments.

(ii) In 1973, international co-operation to protect the oceans led to the Convention for Prevention of Pollution from Ships (MARPOL), whose beneficial effects have been developing since the early 1980s. In the meantime, work was done to begin sketching comprehensive views of the environmental state of the sea [Tracher and Meith-Avcin, 1978].

(iii) Meanwhile, since the mid 1970s, offshore production of oil and gas entered into a rapid growth and diffusion phase: North Sea, Caribbean Sea, Mediterranean Sea, Asian and African marginal seas, and the Arctic Ocean have become theatres of an economic and technological race to exploit submarine fields. Hence there have been three consequences for maritime transportation: (i) the employment of small and medium-sized vessels, as specifically needed by the offshore oil industry, has been implemented; (ii) the network of maritime oil routes has become complicated as a result of the diffusion of areas of exploitation; (iii) the risk of marine accidents has increased because many maritime oil routes pass through straits and choke points.

3.5 OIL AND GAS: THE BIRTH OF OFFSHORE INDUSTRY

The consideration of changes in maritime transportation allow us to take into account the development of the offshore oil and gas industry, which has acquired great importance in some semi-enclosed seas—such as the North Sea [Odell, 1978]. Since the late 1960s, when this industry was born [Symonds, 1978], exploration and exploitation technologies have been progressed unusually quickly. As is well known, moving seawards and extracting hydrocarbons from increasingly deep seabed areas requires extensive development of technologies. In a general sense the 200 m isobath may be regarded as a technological threshold: beyond it, technologies are quite different from, and make much more use of automation than those employed in shallower marine areas. Bearing in mind that advancement beyond this limit has been taking place since the late 1970s, just while offshore production has been spreading to several parts of the oceans, it is clear that a new and significant technological phase is developing.

This process is involving the sea as a whole and for important reasons is concerned with the take off of a new era in the maritime economy.

(i) Although the first generation of technologies—developed till the late 1970s—was conceived in the neo-industrial context, this industrial sector

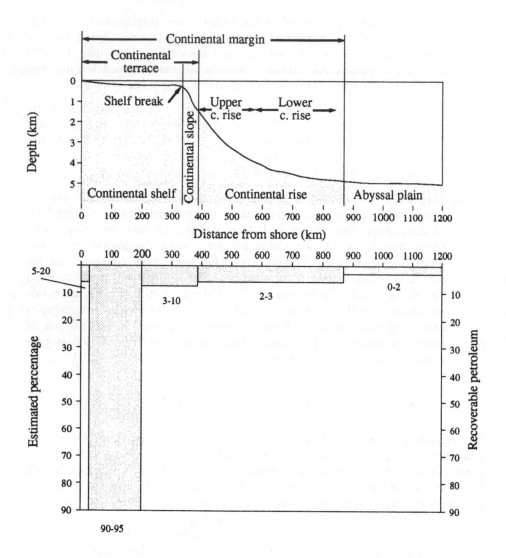

Figure 3.6 *Potential recoverable petroleum in offshore areas.* From Vallega, 1990a, 111, on the basis of data from Post (1983, 11) and Kennett (1982, 27).

is one of the most important for the post-industrial economy, because of the large employment of information systems and robotics and the connections between sea and satellite systems that it requires.

(ii) Thanks to this technological endowment, offshore oil and gas exploration and exploitation provide technologies and organisational models for other uses of the sea, and environmental protection systems as well. Such activities as submarine archaeology, rescue methods, industrial and mercantile platform technologies, submarine pipeline technologies, and marine park establishment, to some extent benefit from the leading role of this industry. As a result, it should be regarded as one of the most interactive and synergistic sea uses.

(iii) On the planetary scale this growing technological patrimony nonetheless makes this industry produce a small environmental impact. Clark [1986, 32] notes that only 0.06 million tons of oil per year flow into the sea from offshore plants.

Focusing attention upon the role of the sea in current global change, the offshore oil and gas industry appears as the first step of the mineral and energy exploitation era. While it is predictable that, for quite a long time, hydrocarbons will be the almost unique mineral resource extracted from beneath the seabed, and that the extraction of minerals from sea water will not advance as much, it is very difficult to predict whether, to what extent and when the exploitation of manganese nodules—the most important resource in the mineral deposits of the deep ocean floor—will take place [UN OETB, 1982; Ford, Niblett, Walker 1987, 7-60]. This is due to two reasons: first, because of the high costs of this kind of seabed exploitation; secondly, because of the environmental impacts that deep-sea mining is able to bring about [Forsund and Strom, 1987]. As a consequence, from the point of view of sea management it is appropriate enough to state that, in short and mid term, oil and gas exploration and exploitation will be the main kind of marine industry, and that for long time only the continental margin will be used for mining. Also the exploitation of non-renewable sources is facing some problems, particularly in areas within national jurisdiction [Brown, 1984].

3.6 CONCLUDING REMARKS

It goes without saying that the transition phase which took place during the 1970s and 1980s is marked also by other significant changes in sea uses and activities [Mann Borgese, 1986, 13-44]. For instance, the exploitation of biological resources is as significant as that of minerals in demonstrating the unusual role of the sea in the current international economy. However, it is not necessary to investigate the whole spectrum of sea uses. Rather, it is important to take into consideration some significant sea uses and processes involving the marine environment—such as those just discussed with the aim of focusing upon methodological concerns. Moreover, the elements that have just been emphasised suffice to demonstrate that this current transition phase is marked by a spectrum of changes that are of great weight for sea management:

(i) the population of many coastal areas is growing more quickly than that of inland regions resulting in very strong human pressure;

(ii) in many parts of the world shorelines consist of artificial structures;

(iii) human activities both on shorelines and at sea are influencing both the sedimentary and chemical cycles;

(iv) the passage from a single or a few use-based management patterns to a multiple use pattern is spreading out, particularly in coastal areas belonging to developed and newly industrialised countries;

(v) the exploitation of marine resources is advancing seawards involving the continental shelf as a whole for oil and gas exploitation and pushing that of biological resources up to and beyond the 200 nm jurisdictional boundary.

This propensity to involve increasingly seaward marine areas in national and corporate strategies is facilitated to varying degrees, according to not only the natural environment in which these take place but also to the legal context from which they can benefit. As a result, attention may now be turned to the relationships between these legal and natural dimensions.

CHAPTER 4

LEGAL FRAMEWORKS AND THE PHYSICAL ENVIRONMENT

4.1 THE CLAIMING OF JURISDICTIONAL ZONES

Sea management requires that special attention be given to the relationships between the legal and physical dimensions. This subject has gained importance in recent decades—particularly since the mid 1960s—because a rapid advance in claiming jurisdictional zones has occurred. According to R.W. Smith's analysis [1986], during the 1958-64 period, i.e. after the first UNCLOS, there were 88 national claims. During the 1965-1971 period there were 160 national claims. Afterwards, when both converging and conflicting interests and political strategies had grown and have been implemented in the context of the Third UNCLOS, national claims reached a peak. During the 1972-1985 period—from the eve of the Conference to the years in which the opportunity of ratifying it was debated—344 national jurisdictional zones were claimed. As a result, there were 592 national claims by the mid 1980s.

According to the data provided by the UN OALOS [Law of the Sea Bulletin, 1986, 8], in 1986 there were 102 200 nm jurisdictional zones; 200 nm broad territorial seas had been claimed by 13 states; Exclusive Economic Zones by 69 states; and 200 nm fishery zones by 20 states. R.W. Smith stresses that "as a result, about 28.5 million sq.nm of ocean space were placed under these coastal states' jurisdiction". In recent years 200 nm national jurisdictional zones have evolved in the way shown by this table.

TABLE 4.1
National jurisdictional zones

200 nautical miles jurisdictional zones	number of states	
	1986	1989
territorial sea	13	12
Exclusive Economic Zone	69	79
fishery zone	20	16
	102	107

From UN OALOS, *Law of the Sea Bulletin*, 1986 (No. 8) and 1990 (No. 15).

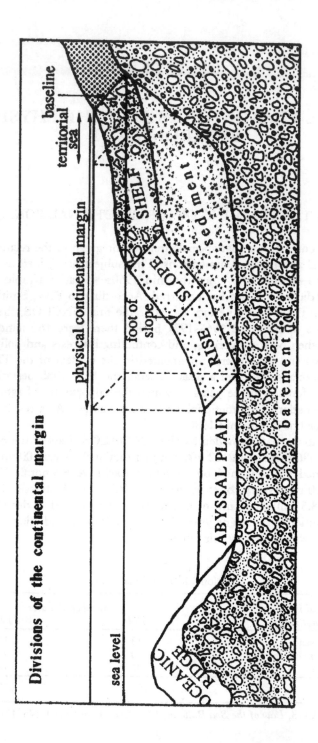

Figure 4.1 *The natural components of a typical continental margin.*

In this context—R.W Smith comments [*Ibid.*], "national claims to the ocean can take two forms: (a) geographical, that is, claims to increased areas; (b) functional, that is, claims to greater regulatory powers within the same geographic area". The evolution of the delimitation of maritime boundaries has brought about a regional assessment of the sea very different from the previous one [Blake, 1987; Smith, 1987]. It can be stated that the regionalization of the sea has come to light above all because of the setting up of jurisdictional zones.

At this point the complexity of legal frameworks is self evident and seems to advance as national policies based on opportunities provided by the 1982 UN Convention on the Law of the Sea bear fruit. Moving seawards, states may in their policies make claims and establish regulations for internal waters, the territorial sea, the contiguous zone, the continental shelf and Exclusive Economic Zone—or, alternatively, the Exclusive Fishery Zone (Figures 4.1 and 4.2). Beyond the outer limit of the largest national jurisdictional area—e.g., the Exclusive Economic Zone—there extends the international régime: high seas and the deep seabed.

On the other hand, management complexity is not smaller than that of the legal context. Moving seawards three areas may be discussed:

(i) the *urban and extra-urban waterfront*, the importance of which has been growing since the early 1970s;

(ii) the *coastal area*, to be evaluated both in its current and prospective extents, the latter involving the continental margin as a whole;

(iii) the *ocean area*, namely, the marine area extending beyond the outer edge of the continental margin and including islands and archipelagic areas.

Legal complexity is here worth considering, putting national and international zones in relation to natural environments, while management complexity will be dealt with in the next chapter.

4.2 INTERNAL WATERS AND TERRITORIAL SEAS

Waters extending landward from baselines possess three main features. From the *natural* point of view they are concerned with the most fragile coastal environment because they are largely involved in erosion processes, are affected by the interaction between salt and fresh waters and are endowed with such fragile ecosystems that they need extra protection. From the *legal* stand-point they are the ground where coastal, island and archipelagic states have their entire sovereignty to the point that they are regarded as a real "territorial space" [Santis Arenas, 1988]. From the *organisational* stand-point they are strongly involved in urban and extra-urban waterfronts so they are becoming the foci of present patterns of coastal management.

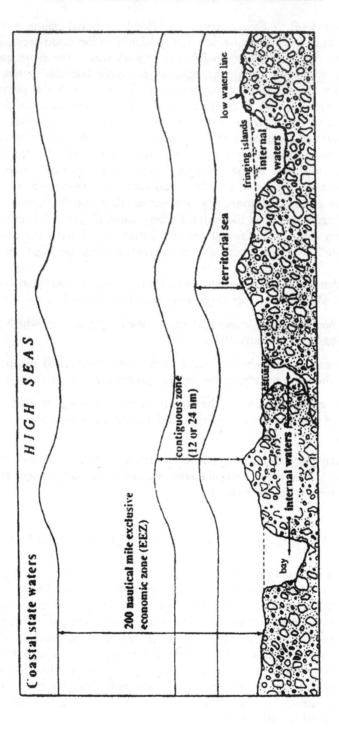

Figure 4.2 *The general scheme of the maritime national zones (except the continental shelf).*

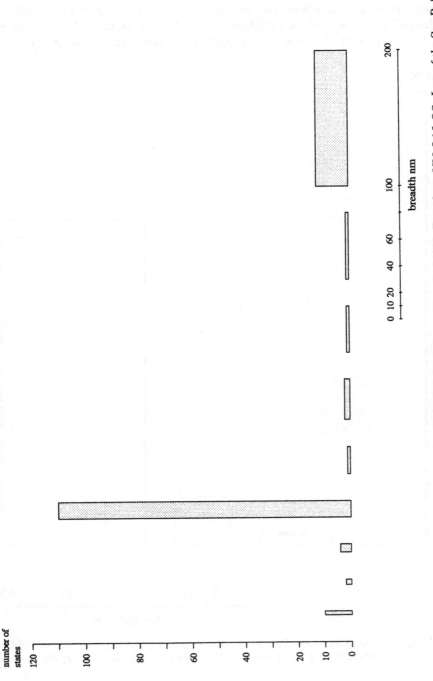

Figure 4.3 *The territorial seas according to their breadth claimed up to the end of 1989. Data from UN OALOS, Law of the Sea Bulletin, 1990, No. 15, page 39.*

When the Third UNCLOS started, most states were endowed with a 3 nm territorial sea and a number of states had claimed—or were about to claim— 12 nm territorial seas. Since that time this situation has largely been realised. As defined in Article 3 of the 1982 Convention on the Law of the Sea, 110 states (76%) have claimed 12 nm territorial seas. More extended territorial seas have

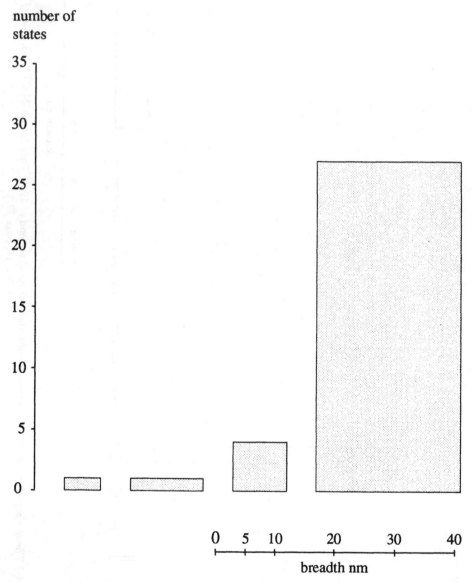

Figure 4.4 *The contiguous zones according to their breadth* claimed up to the end of 1989. Data from UN OALOS, *Law of the Sea Bulletin*, 1990, No. 15, page 39.

been claimed by 17 states (12%), 11 of which have prolonged this jurisdictional area up to 200 nm breadth giving birth to an important legal issue (Figure 4.3).

As is well known, to some extent the 1982 UN Convention (Articles 7 to 16) was innovatory in providing the criteria by which straight baselines are defined. After having established that "the baselines for measuring the breadth of the territorial sea is the low-water line along the coast as marked on large-scale charts officially recognised by the coastal state" (Art 5), the Convention takes into consideration reefs, mouths of rivers, bays, seaports, roadsteads, and other features, providing a detailed list of criteria which in general allow the state to establish its baselines more to seaward than in the past. Bearing in mind what important benefits the state is able to draw from the territorial sea, the ability to establish baselines as far to seaward as is admissible (in some cases even farther) in the Convention has spread. By moving baselines seaward two advantages are realised.

(i)   *Minimisation of constraints*. Internal waters reach their largest breadth compatible with the international law of the sea. As a result, the state benefits from the largest space in which it is capable of exercising complete sovereignty, without being affected either by constraints imposed by the international law of the sea or being disturbed by claims from opposite or adjacent states. Moving from internal waters to the territorial sea the content of national jurisdiction undergoes only small changes concerning navigation (aliens have the right of innocent passage), overflight (aliens can overfly straits used for international navigation) and scientific research [Prescott, 1985, 40-1]. In conclusion, both the internal waters and the territorial sea provide the best jurisdictional basis for managing the waterfront and the coastal area.

(ii) *Comprehensive management*. In both internal and territorial waters the state is able to set up plans and subsequent management involving the marine environment in its entirety: sea surface, water column, seabed and subsoil. In addition, it has enough legal tools to manage the shoreline and the sea in an integrated way. Zones ranging from the land to the outer limit of the territorial sea may be regarded as the area in which a comprehensive approach to management may be attempted.

(iii) *Environmental protection and preservation*. The state has complete authority to make regulations concerning environmental protection and to control human presence and activities in internal and territorial waters.

4.3 THE CONTINENTAL SHELF

Moving seaward from the outer limit of the territorial sea the complexity of the legal framework—as it is evolving according to the 1982 Convention—increases and gives impetus to the possibilities of managing marine areas. To a large extent this complexity is due to the fact that the law of the sea, in its historical development, has constructed two zones—the continental shelf and the Exclusive Economic Zone—both pertaining to the 200 nm zone but very differently affecting the potential for managing marine resources and protecting ecosystems. With reference to sea management some implications from these legal patterns are worth pointing out.

As far as the continental shelf is concerned, the first international agreement— i.e. that from the 1958 UNCLOS—led to limiting the continental shelf "to the seabed and the subsoil of the submarine areas adjacent to the coast but outside the area of the territorial sea, to a depth of 200 metres or, beyond that limit, to where the depth of the superadjacent waters admits of the exploitation of the natural resources of the said area". Through such a concept:

(i)   the coastal state acquires the right to exploit resources anywhere up to the 200 metre isobath, leaving out of consideration the fact that—moving from the continental shelf to the ocean areas—the marine environment and resources change thus requiring different management approaches;

(ii)   a very limited spectrum of management possibilities exists in this jurisdictional context because, by claiming the continental shelf, the coastal, inland and archipelagic state can exploit only seabed and subsoil resources;

(iii) national jurisdiction is not concerned with the protection and preservation of the marine environment *per se*.

This concept of the continental shelf was replaced by the 1982 Convention. According to Art 76.1, the continental shelf is "the prolongation of the land territory to the outer edge of the continental margin, or to a distance of 200 nautical miles from the baseline from which the breadth of the territorial sea is measured". By means of this definition the (legal) continental shelf could cover the entire (physical) continental margin or be prolonged also beyond the outer edge of the continental margin to cover ocean areas. Each case is provided with its own legal discipline. If the outer edge of the (physical) continental margin extends beyond 200 nm from the baseline, the (legal) continental shelf is prolonged to this limit. If the outer edge does not extend up to 200 nm, the continental shelf is considered as extending up to this limit.

As far as the (legal) continental shelf concept is concerned, attention should be paid to the implications arising from the 1982 Convention framework. This new legal approach brought about the passage from a concept based on the depth of the sea—the 200 m isobath was the main reference line—to a concept based on the distance from the baselines. In fact, the new régime refers also to the depth-based criteria but in a secondary way, as can be seen by Art 76.5: "The fixed points comprising the line of the outer limits of the continental shelf on the sea-bed (...) either shall not exceed 350 nautical miles from the baselines from which the breadth of the territorial sea is measured or shall not exceed 100 nautical miles from the 2,500 metre isobath (...)" (Figure 4.5).

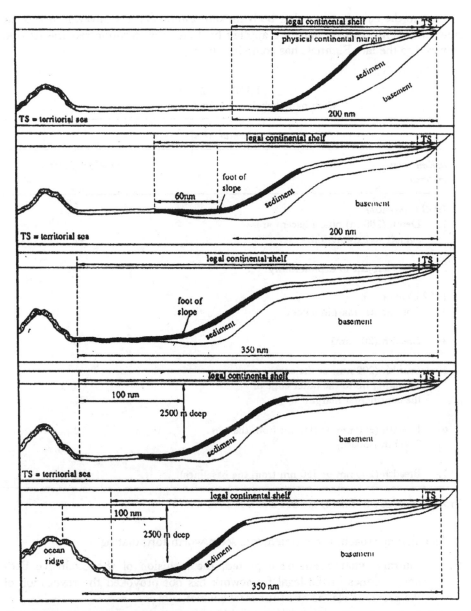

Figure 4.5 *Criteria for delimiting the continental shelf*, based on the approach of *Atlas of the Oceans* (Couper ed., 1983, 223). From above: A) where the continental margin is narrower than 200 nm, the continental (legal) shelf can extend beyond this limit; B) a variation of the so-called Irish formula, according to which the extent of the continental shelf, where it is narrower than the continental margin, is established by a line 60 nm distant from the foot of the continental slope; C) the extent of the continental shelf cannot exceed 350 nm: D) in alternative it cannot extend more than 100 nm from the 2500 m isobath; E) it cannot in anyway encompass ocean ridges.

Due to the circumstance that both the old criteria (1958 Convention) and the new ones (1982 Convention) are applied by states this complicated framework, referred to the late Eighties, has come into being.

TABLE 4.2
Continental shelves: delimitation criteria

continental shelf	
delimitation criteria	number of states
1958 Convention	
1. Depth (200 m) plus adjacent areas subject to exploitability	42
2. Areas subject to exploitability	4
1982 Convention	
3. Continental margin extent	1
4. Breadth (200 nm)	6
5. Breadth (200 nm) or the area extending up to the outer edge of the continental margin	21
6. Breadth (200 nm or 100 nm from the 2,500 m isobath)	2
7. Breadth (200 nm or 350 nm from the baselines)	1

From UN OALOS, *Law of the Sea Bulletin*, 1990, No. 15.

As a first approach, three statements seem worth formulating:

(i)   contrary what might be expected, the inclusion of the Exclusive Economic Zones in the legal framework has not provoked the revocation of the previous claims of continental shelves.

(ii)   in the present context both 1958 and 1982 criteria coexist in such a way as to provide a very large spectrum of kinds of continental shelves claimed on the basis of depth and breadth;

(iii) in many semi-enclosed seas, continental shelves survive, and Exclusive Economic Zones have not been claimed, which is one of the most intriguing features of the present stage of sea management.

4.4 THE EXCLUSIVE ECONOMIC ZONE

There is no doubt that, in general, the Exclusive Economic Zone, as established by the 1982 Convention, is much more consistent with the need to manage the sea than the continental shelf is (Figure 4.6). As a matter of fact, it is regarded as an area beyond and adjacent to the territorial sea in which the coastal state has sovereign rights for tasks including, exploring and exploiting, conserving and managing the natural resources, for placing and using artificial islands and structures, developing scientific research, and protecting and conserving the marine environment (Art. 56). The delimitation is relevant to both mineral [Earney, 1987] and biological resources [Langford, 1987; Wise, 1987] exploitation. The Exclusive Economic Zone may be claimed by he coastal, island and archipelagic state or agreed between states with opposite or adjacent coasts when the distance between their baselines is less than 400 nm. Because of this, the claiming of Exclusive Economic Zones spread during and after UNCLOS III and has become the most intriguing process involving the sea and also the main tool for extending the space in which national interests can be pursued.

A latere of the Exclusive Economic Zone the Exclusive Fishery Zone should be considered because of its importance both for international law [Brown and Churchill, 1985] and national policies. In the late Eighties 20 states had claimed and 5 states had agreed their fishery zones (Figure 4.7). Most of these zones extend 200 nm so, as a matter of fact, they are a sort of Exclusive Economic Zone where rights of the coastal, inland and archipelagic states are presently claimed only for the exploitation of biological resources—regardless of whether the continental shelf has been claimed or agreed. There follows a list of fishery zones according to their breadth.

TABLE 4.3
The breadth of fishery zones

breadth (nm)	fishery zones number of states
12	2
25	1
50	1
200	16
up to the median line with neighbouring states	5

From UN OALOS, *The Law of the Sea Bulletin*, 1990, No. 15. Reference: 1989.

Comparing Articles 76 and 83, on the delimitation of the continental shelf, and Art 74, concerning the Exclusive Economic Zone, it appears that the extant concept of the latter jurisdictional zone is less complex than that of the former. For the Exclusive Economic Zone the limit of 200 nm from the baselines is a

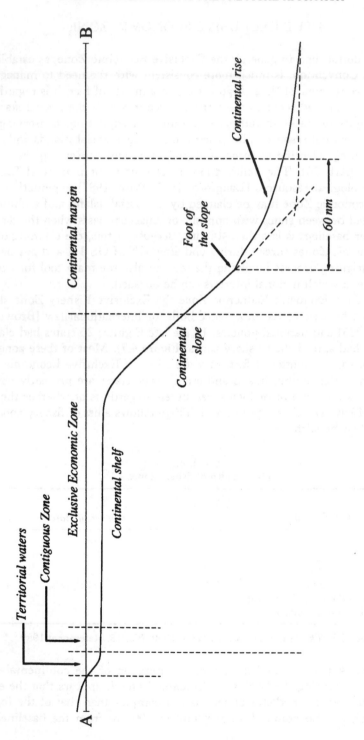

Figure 4.6 *The profile of national jurisdictional zones (except the continental shelf) according to the natural environment. From Prescott, 1985, 37.*

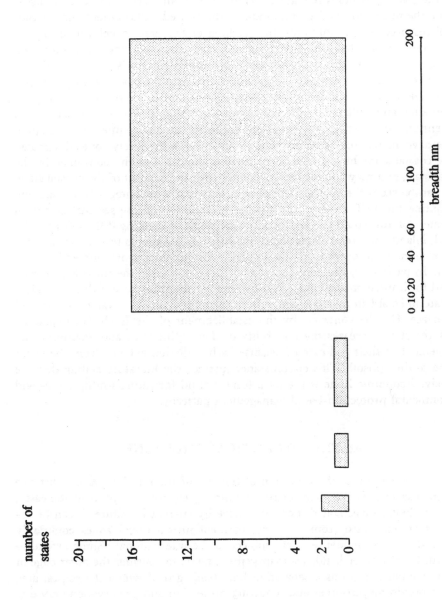

Figure 4.7 *The fishery zones according to their breadth* claimed up to the end of 1989. Data from UN OALOS, *Law of the Sea Bulletin*, 1990, No. 15, page 40.

rigid outer boundary while, as far as the continental shelf is concerned, this limit should be regarded only as a reference line. In addition, the continental shelf enables the state to exploit the seabed and the subsoil and, according to it, to legislate for the protection and preservation of the seabed environment. As for biological resources it can only limit the catching of sedentary species by aliens. In this context aliens have a wide range of rights—concerning navigation and overflight, fishing and research. But different implications arise when the Exclusive Economic Zone is claimed or agreed. In this case the coastal state acquires a wide range of rights concerning artificial islands and offshore installations, the conservation and utilization of living resources, research (Arts 60-68 of the 1982 Convention) and, lastly, has complete authority in environmental legislation. Aliens have navigation rights providing they observe the safety zones designated by the coastal state, have access to the surplus allowable catch determined by the coastal state, and may conduct research only with the consent of the coastal state.

As can be seen, the coastal state derives only a few advantages from claiming the continental shelf. In this case, the aliens' activities on the sea surface and in the water column could be developed to the point of creating difficulties for the coastal, inland and archipelagic states willing to give impetus to sea management both in terms of resource exploitation and environmental protection and preservation. In particular, the continental shelf does not seem the most appropriate ground for implementing comprehensive management because it does not allow the state to establish management systems also involving the water column and sea surface. On the contrary, by the establishment of the Exclusive Economic Zones the state increases the possibility of developing uses and environmental management to their full extent, because its jurisdiction extends from the water surface to the subsoil. This circumstance justifies the literature stating that the Exclusive Economic Zone is the focal legal ground for future multiple use- and environmental protection-based management patterns.

4.5 THE 200 NAUTICAL MILE LINE

At the present stage of the evolution of the law of the sea the 200 nm distance from the baselines is the main reference limit. Apart from the problematic cases, which are being investigated more and more by current literature, it can be assumed that, landward from this line, national jurisdictional zones commonly occur and, as a consequence, areas subject to the international régime extend to seaward. On the other hand, sea management tends to consider the outer edge of the continental margin as a separation line, landward of which the coastal area extends providing resources and requiring protection and preservation very different from those of the ocean area.

If the outer limit of the largest national zone—namely, the limit of the continental shelf or the Exclusive Economic Zone—coincides with the outer edge of the continental margin, the rationale occurs because both the legal and natural contexts have the same extent and the national jurisdictional space would co-incide with the coastal area. Beyond these limits the marine spaces subject to the international régime and the ocean area extend. As a consequence, in order to provide ground for sea management the relationships between the extent of the continental margin, on the one hand, and those of the (legal) continental shelf and Exclusive Economic Zone, on the other, are to be taken into account.

Bearing in mind that, in the late Eighties, states had claimed or agreed their continental shelves, the first step of analysis should be concerned with the relationships between the extent of such a jurisdictional zone and the (physical) continental margin. Due to the fact that current continental shelves are partly based on the figure provided by the 1958 Convention and partly on that provided by the 1982 Convention, at the present time the extent of the continental (legal) shelf can coincide with that of the continental margin or be narrower or wider than this. As a consequence, when

(i)   the continental (legal) shelf has the same extent as the continental margin, the national jurisdictional space relates only to the coastal area and the state can perform only coastal management;

(ii) the continental (legal) shelf is narrower than that of the continental margin, the national jurisdictional space is less extended than the coastal area and the state can perform only coastal management;

(iii) the continental (legal) shelf is wider than the continental margin, it includes both the coastal area and an ocean space and theoretically it enables the state to develop both coastal and ocean management.

Of course, coastal management supported by the continental (legal) shelf is partial because it concerns only the seabed and subsoil. On the contrary, management supported by the Exclusive Economic Zone is comprehensive because the marine context as a whole can be exploited and protected. Bearing in mind that this zone extends 200 nm from the baselines whichever is the breadth of the continental margin, the three cases mentioned occur again: the breadth of the Exclusive Economic Zone may be wider or narrower than that of the continental margin or—when a border-line case takes place—can coincide with this.

Since states can contextually be provided with the continental shelf and the Exclusive Economic Zone, sea management has to take into account both the extents of these jurisdictional zones and, on the other hand, the breadth of the continental margin. A complicated framework follows, which will be considered in Chapter 6. Here it is useful to briefly survey the breadth of the continental margin in various ocean areas in relation to the 200 nm line (Figure 4.8).

Polar seas: Arctic Ocean. In Arctic areas margins are Atlantic, i.e. divergent and aseismic. Except in a few areas, the 200 nm line crosses the continental margin.

Polar seas: Antarctic Ocean. The Antarctic margins are also divergent and aseismic. Most of them consist of a narrow shelf and quite wide slopes and rises. The 200 nm line crosses the slope or rise except in some areas, such as in the western part of the Weddell Sea, where it is located beyond the outer edge of the continental margin. Because it was agreed (1991) that the Antarctic area is to be protected and preserved, the role of the 200 nm line is not concerned with current patterns of coastal area management because resource exploitation is not be possible.

Eastern Atlantic Ocean. As a first approach, the eastern side of the Atlantic Ocean, consisting of divergent and aseismic margins, can be divided into several parts.

(i)   The *Norwegian Sea*: almost everywhere the outer edge of the continental margin extends less than 200 nm; the shelf is wide, slopes and rises are quite narrow.

(ii) The *North Sea and the Atlantic ocean surrounding the British archipelago*: the 200 nm line almost everywhere crosses the continental margin.

(iii) The *Atlantic waters of Morocco and western Africa*: the (physical) continental shelf is narrow but slopes and rises are wide so the outer edge of the continental margin extends almost everywhere beyond the 200 nm line.

(iv) *Western and central part of the Gulf of Guinea*: the 200 nm line tends to coincide with the outer edge of the continental margin.

(v)   *Eastern side of the Gulf of Guinea and the Atlantic waters as far as South Africa*: the outer edge of the continental margin runs seaward of the 200 nm line. The continental (physical) shelf is narrow.

Western Atlantic Ocean. The northern and southern side of Iceland, the southern side of Greenland and the eastern side of the Bahamas and Lesser Antilles are provided with a continental margin not extending up to 200 nm. In the rest of this Atlantic side the continental margin is wider than 200 nm, the continental (physical) shelf is narrow along the northern and central coasts of the South America and wide (more than 200 nm) at the latitude of Newfoundland and along the Argentinian coast.

Indian Ocean. The assessment of this ocean is so complicated as to need detailed analysis. Its margin is of Atlantic type, i.e. divergent and aseismic, except on the north-eastern side—from the eastern side of the Gulf of Bengal to the northern and western Australian coasts—where it is of the Pacific type, i.e. convergent and seismic.

The continental margin is wider than 200 nm in the central section of the Arabian Sea, in the central and eastern areas of the Gulf of Bengal, along East African and West Australian coasts. Except in Australia and in the Andaman Sea, the continental (physical) shelf is narrow so the continental margin consists above all of the slopes and rises.

Eastern Pacific Ocean. Its margins are convergent and seismic and are narrower than 200 nm except in a zone extending approximately from the Mendocino Fracture Zone up to the Gulf of Alaska. In the rest of the waters surrounding most of the Northern, Central and Southern American coasts the (physical) continental shelf is narrow almost everywhere and slopes and rises are almost non-existent. As a consequence, the 200 nm line runs seaward far beyond the outer edge of the continental margin.

Central and Western Pacific Ocean. This oceanic area provides the most complicated situation in the world [Buchholz, 1987]. From the northern limits to Japan the continental margin is narrower than 200 nm but its (physical) shelf is quite wide. Micronesia and Polynesia consist of oceanic islands so they do not have a continental margin. New Zealand is provided with a wide (physical) continental shelf except on the north-eastern side of North Island and in the south-western side of South Island. Slopes and rises are wide on the western sides of both islands. As a result, the 200 nm line crosses the continental margin except in the south-western and in the southern sides.

Semi-enclosed seas. At the present time these seas probably afford the most interesting issue for sea management. This is due to at least three features. First, it depends on the complexity of the natural environment because only a few (such as the North Sea) consist of continental shelf. On the contrary, most—such as the Mediterranean Sea [Kliot, 1987], the Caribbean Sea and regional seas of South-East Asia—also have geomorphologically complex slopes and rises, and sometimes abyssal plains as well and are seismic. Secondly, the seas located in the intertropical and temperate zones are affected by high human pressure generated by a wide spectrum of uses, so that both resource and environmental management face considerable difficulties. Thirdly, their management requires appropriate legal frameworks based on international agreements and inter-regional co-operation, as the Mediterranean Action Plan's experience has demonstrated [Raftopoulos, 1988]. In particular, almost everywhere both the continental (legal) shelves and Exclusive Economic Zones cannot be established only through agreements between opposite and contiguous states.

4.6 BEYOND THE NATIONAL JURISDICTIONAL ZONES

At the present stage of reasoning it is not worth considering whether a state has claimed or agreed its continental shelf, enabling it to make only partial sea management, or the Exclusive Economic Zone, enabling it to develop comprehensive management: this issue will be faced in Chapters 5 and 6. Rather, it is now appropriate to consider the wide spectrum of areas which arises from the preliminary review of relationships between the breadth of the continental margin and the position of the 200 nm limit. It demonstrates that the possibility of making the main national jurisdictional zone (continental shelf or Exclusive Economic Zone) coincident with the extent of the continental margin (or, at least, of the physical continental shelf) is rare. The interplay between the legal and physical frameworks provide an unusual ground for the management of resources and natural environment.

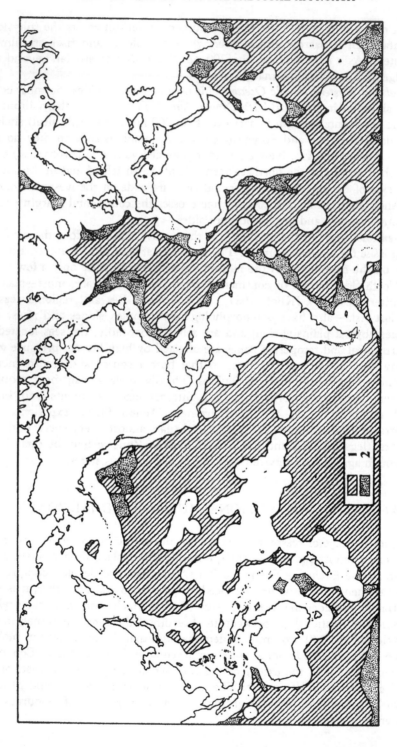

Figure 4.8 *The extent of the continental margin vis-à-vis the 200 nm line.* From Prescott, 1985, 18. 1) areas over deep seabed more than 200 nm from territory; 2) margins more than 200 nm from territory.

According to Prescott [1985, 96], the most interesting aspect of this problematic issue occurs where the states, which do not have a jurisdictional zone extending up to the outer edge of the continental margin, tends to prolong it up to this limit. It should be regarded as an effort to develop a rationale in sea management. Prescott [*ibid.*] points out that the prolongation of the jurisdictional zone is pursued according to these three complementary criteria:

(i)   The *gradient of the seabed.* "Since the general tendency of continental margins is to decline in elevation from the coast to the abyssal plain, some states might argue that so long as the gradient is downwards from their shore, a claim to the margin should succeed".

TABLE 4.4
A classification of continental margins

Only one country is involved
  1 The margin is narrower than 200 nm
  2 The margin is wider than 200 nm

Two opposite countries are involved
  1 The countries are less than 400 nm apart
    1.1 The margins are separated by the deep seabed
      1.1.1 Neither margin extends across the median line
      1.1.2 One margin extends across the median line

    1.2 There is a continuous margin between the two countries
      1.2.1 The margin is undifferentiated
      1.2.2 The margin is differentiated by a zone which does
            not include the median line

  2 The countries are more than 400 nm apart
    2.1 The margins are separated by the deep seabed
    2.2 There is a continuous margin between the two countries
      2.2.1 The margin is undifferentiated
      2.2.2 The margin is differentiated by a zone which does
            not coincide with the median line

Two adjacent countries are involved
  1 The margin is narrower than 200 nm
    1.1 The margin is undifferentiated
    1.2 The margin is differentiated by a zone which does not
        include the median line

  2 The margin is wider than 200 nm
    2.1 The margin is undifferentiated
    2.2 The margin is differentiated by a zone which does not
        include the median line
    2.3 A spur from the continental margin of one country
        passes in front of the continental margin of its
        neighbour

From Prescott [1985, 97].

(ii)   The *structure of the seabed*. It should be "investigated to discover whether is was formed and deformed by the same processes and forces which moulded the coastal regions".

(iii)   The *geological features of the coast and seabed*. They should be compared to reveal whether there are continuities between the two zones.

Prescott concludes that these issues are "a matter for speculation, which would probably never be conclusive, how much weight should be attached to each of these characteristics, or to combinations of them" [*ibid.*].

Up till now the case of a single state facing the marine area has been considered. More complicated situations occur when two countries are opposite or adjacent: a range of implications arises according to whether the distance between their coasts is narrower or wider than 400 nm. First of all, complexity is legal but it also deeply involves management patterns. This is self-evident when the classification of margins formulated by Prescott [1985, 97] is considered (Table 4.4).

As can be seen, where two countries are opposite seven hypotheses occur and where they are adjacent five hypotheses appear: each of these brings about peculiar implications in sea management.

4.7 OCEAN AREAS AS INTERNATIONAL ZONES

Approaching the international legal context different kinds of areas are encountered. According to the 1982 Convention these consist of the high seas and the Area, i.e. the deep-sea beds. As far as these international areas extend only beyond the outer limit of the (physical) continental margin they involve ocean ridges and rises, ocean-basin floor and continental rises, are concerned with most of mineralisation processes and with the main surface and deep currents. The natural environment here is surely less complicated than the continental margin but it is ground for the main processes involving the ocean, and is provided with a great patrimony of ecosystems that, according to the latitude, consists of food webs of varying degrees of complexity. In general, areas subject to the international régime are less difficult to manage than coastal areas, due both to human pressure, which is weaker, and the current level of technology, which does not provide tools to appropriately and economically exploit ocean resources.

Subject 4.1
The teleology of the common heritage of mankind

"The effort to implement the principles of equity and of common heritage of mankind in the seas was the major impulse in the decision of the United National General Assembly to convene the (...) Law of the Sea Conference. This effort gave a focus to the early stages of the

negotiations. If the effort is abandoned and if the principles are for-
gotten, the Conference cannot achieve enduring results, even if
agreement on a treaty is reached. The present challenge requires
looking beyond the immediate and the obvious so as to ensure the
necessary protection of national interests within a framework which
is recognized as fair; which takes into account technological, economic, and environmental factors; and which flexibility reconciles a
global order with regional needs." [Pardo, 1978, 34].

As is well known, from the legal point of view the idea of the high seas has
passed through great changes. In the 1958 Convention a simple concept was formulated: high seas were defined as the marine space extending beyond the territorial sea. Hence five consequences:

(i)   the outer limit of the territorial sea of coastal, inland and archipelagic
      countries marked the border between the national jurisdictional area and
      the international space;

(ii) the management of the sea was based on a very simple legal framework
      and took place in a narrow zone largely from 3 to 6 nm depending on the
      extent of territorial seas;

(iii) the extra-national sea was conceptually uncomplicated since it was deter-
      mined only in terms of high seas;

(iv) the extent of high seas was very wide since most of the ocean space be-
      longed the international régime;

(v)   high seas covered also large parts of the continental margin.

This assessment was due to the fact that national policies were still not oriented towards implementing the exploitation of marine resources and there were
a few conflicts between coastal, island and archipelagic states. In addition, the
spectrum of sea uses was not wide. At that time transportation, fishing and defence were being developed, larger and larger littoral industrial areas were being
built and greater and greater flows of raw materials were crossing the oceans.
But the offshore oil and gas industry was still in a pre-take off phase, deep-sea
mining had still not emerged and the comprehensive exploitation of coastal areas
was far from being realised.

During the second half of the 1970s and in the early 1980s the economic and
social contexts changed: (i) the offshore industrial technologies benefited from
impressive improvements and were entering a take off phase; (ii) in particular,
offshore oil and gas industry was regarded as a leading activity; (iii) the
prospects of drawing energy from the sea seemed promising; (iv) deepsea mining
was regarded as a new industrial frontier to be set up in the short term; (v) aquaculture was progressing to a significant degree; (vi) new activities at sea, such as
archaeology, were developing; (vii) the prospect of creating undersea settlements

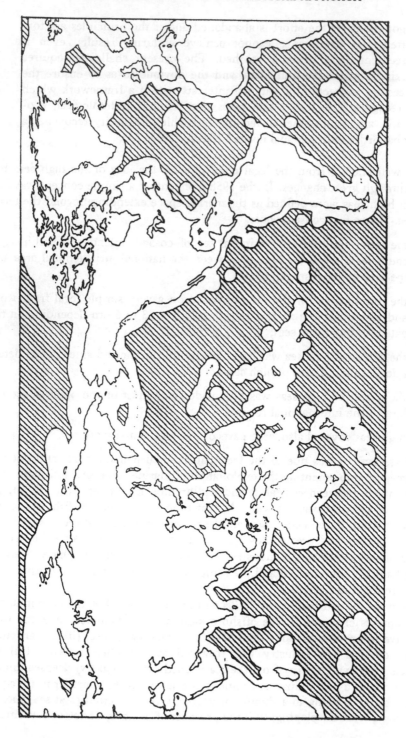

Figure 4.9 *The extent of the high seas.* From Prescott, 1985, 110.

was in the offing. Hence two consequences: on the one hand, the extra-national maritime areas were drastically reduced because of the establishment of the Exclusive Economic Zone, the change in criteria for delimiting baselines and the enlargement of the breadth of the territorial sea; on the other hand, the international régime became more complicated since the category of high seas was flanked by that of deep-sea beds. This process is worthy of a *coup d'oeil*.

As far as sea management is concerned, it should first be considered that, according to Art. 86 of the 1982 Convention, high seas concern "all parts of the sea that are not included in the Exclusive Economic Zone, in the territorial sea or in the internal waters of a State, or in the archipelagic waters of an archipelagic state". As can be seen, three implications arise: (i) in most parts of the world the Exclusive Economic Zone is the real current jurisdictional zone to which the extent of high seas is to be referred; (ii) the (legal) continental shelf is not consistent with the delimitation of high seas although somewhere this zone is extended beyond the outer edge of the Exclusive Economic Zone; (iii) archipelagic waters, when they belong to an archipelagic state, are not to be considered as high seas. Both the archipelagic state and archipelago are defined by Art. 46 of the 1982 Convention: the former means "a state constituted wholly by one or more archipelagoes and may include other islands"; the latter "means a group of islands, including parts of islands, interconnecting waters and other natural features which are so closely interrelated that such islands, waters and other natural features form an intrinsic geographical, economic and political entity, or which historically have been regarded as such".

As a consequence, as Prescott [1985, 109-113] demonstrated, the extent of high seas has been reduced to sixteen areas but only two are very large (Figure 4.9). "The largest consists of vast expanses of the Atlantic, Pacific and Indian Ocean, which are connected south of Africa, Australia and South America. While the connecting passages south of Australia and Africa are wide, that which lies south of South America would be restricted to a width of 40 nautical miles if national claims were permitted from Antarctica". The second large area is found in the Arctic Ocean.

Subject 4.2
Small areas of high seas

Small areas of high seas "The remaining fourteen areas of high seas are also enclaves surrounded by national waters, but they are all much smaller than the high seas of the Arctic Ocean. Only four of these small enclaves of high seas are found outside the Pacific Ocean. In the Norwegian Sea, between the Norwegian mainland and Iceland, Jan Mayen and Greenland there is an elongated area of high seas (...). To the east of this areas a smaller enclave of high seas is located in the Barents Sea between Svalbard and Novaya Zemlya (...). Two very small enclaves of high seas are found in the Gulf of Mexico

(...). The ten enclaves of high seas in the Pacific Ocean are all on the western margin. (...). To conclude this section describing the distribution of high seas it is only necessary to record that high seas have been totally eliminated in all the enclosed and semienclosed seas apart from the Arctic Ocean and the Gulf of Mexico". [Prescott, 1985, 111-113].

According to Art. 87 of the 1982 Convention, the high seas are the realm of the freedom at sea. *Inter alia*, freedom includes navigation, overflight, submarine cables and pipelines, the construction of artificial islands and other installations, fishing and scientific research. The state exercising these freedoms has to regard only the interests of other states in the exercise of the freedom of the high seas.

While the legal concept of high seas has ancient roots in history and juridical culture, that of deep seabeds is new and it is be regarded as a peculiar product of post-industrial society. The legal precedent can be found in the well known resolution no. 2749-XXV adopted by the UN General Assembly in 1970 on the deep-sea beds as a "common heritage of mankind"—the *expression frappante* that was included in the 1982 Convention (Art. 136). This category has appeared in the 1982 Convention and was named as *Area* and—this is worth stressing—was not defined. Art. 133 establishes what should be regarded as "resources" and points out that "when they are recovered from the Area they are referred to as minerals". In conclusion, the Area consists of seabeds and subsoil extending beyond zones subject to national jurisdiction.

Subject 4.3
The Area: terms and principles from the 1982 Convention

Art. 133: "'resources' means all solid, liquid and gaseous mineral resources *in situ* in the Areas at or beneath the sea-bed, including polymetallic nodules; resources, when recovered from the Area, are referred to as 'minerals'".

Art. 135. "Neither this Part nor any rights granted or exercised pursuant thereto shall affect the legal status of the waters superadjacent to the Area or that of the air space above those waters."

Art. 137.3 "No State or natural or juridical person shall claim, acquire or exercise rights with respect to the minerals recovered from the Area except in accordance with this Part. Otherwise, no such claim, acquisition or exercise of such rights shall be recognized."

As Prescott observes [1985, 125], the sixteen areas of high seas do not overlie the deepsea beds since—as is has mentioned—somewhere, such as in marine spaces surrounding the Soviet Union and New Zealand, high seas overlie the (legal) continental shelf, while the deep seabed must extend beyond the national jurisdictional zones involving the seabed and subsoil. As a consequence, only "ten of the areas of high seas will have corresponding areas of deep seabed, but only in three cases will the limits of the high seas and deep seabed correspond exactly". The largest area of the deep seabeds extends from the Pacific to the Atlantic through the Indian Ocean, although it is not continuous south of the South America [*ibid.*].

The Area was included in the legal framework under the pressure of policies oriented to the exploitation of manganese deposits and other mineral resources of the deep seabed and subsoil. This prospect activity—whatever term it is to be referred to—is not the unique issue of ocean management. Also conventional sea uses, such as navigation, fishing and submarine cables, concern it. Lastly, the need for the protection and preservation of the ocean environment acquires more and more importance in management. That justifies stating that the relationships between the legal framework and the natural environment require analyses on the regional scale, as the coastal areas do.

CHAPTER 5

THE SEA USE STRUCTURE

5.1 STARTING POINTS

Consideration of sea management is, as a general rule, a first step towards taking into account the specific management of coastal and ocean areas. To this end a set of statements from Chapter 1 is worth recalling. The evolution of national policies and international co-operation [Smith H.D., 1988] towards the sea has led to the setting up of more and more complicated marine management structures consisting of the natural environment and sea uses: recently the former component has increasingly been jeopardized; the latter has rapidly evolved and has become more and more complicated. The evolution of sea uses has passed through phases in which technologies, organisational patterns and management strategies have changed so radically as to transform the sea use structures of many areas. These morphogeneses have been affected by changes in the web of relationships among uses, as well as in the interactions between uses and the environment. The most fundamental changes have occurred since the 1970s.

According to general system-based theory the change in sea management (adjustments or morphogeneses) can be explained (i) starting from the analysis of the sea use structure, (ii) considering the process which the sea use structure undergoes, (iii) identifying the goals to which the process leads and (iv) subsequent changes involving the sea use structure. As can be seen, the investigation follows a circular path.

Sea management is a realm of complexity, not only because more and more complicated sea use structures are being created, but also because scientific thought and research have contributed to the creation of complicated reasoning patterns and investigative models with the aim of describing the sea use structure, particularly when it passes through phases of change. Complexity of thought reflects complexity of reality.

This is especially due to the circumstance that sea management brings about the interpenetration of two environments—social and natural—that are self-poietic since both of them are capable of producing their own organisational elements and internal reassessments, and addressing of outputs to the external environment. From the methodological point of view it follows that such environments cannot be described in terms of cause-effect relationships nor, in a general sense, through Cartesian logic. The interaction between them has a non-deterministic nature and structures created by this interaction are not trivial machines since they are able to make a range of outputs from the same input.

5.2 THE SEA USE STRUCTURE AS A WHOLE: THE SEA USE FRAMEWORK

As a first approach, the sea use structure can be considered as a set of natural and human elements, such facilities and activities at sea, and resource use, which are more closely tied each other than they are to other elements which belong to the external environment [Kenchington, 1990, 28-59]. This set consists of two sub-sets of elements, natural and social: the former consists of the ecosystem and its physical background; the latter embraces sea uses and human activities. According to general system theory the use structure, when considered in relation to its path in time—i.e. in relation to its evolution—is a system. As a result, it is poss-ible to state that sea management gives birth to sea use structures and subsequent sea systems. As far as the human presence on the sea is concerned, current liter-ature tends to regard sea uses as consisting of the exploitation of the marine en-vironment involving the consumption of matter and energy, while other marine activities—e.g. navigation, parks and reserves—do not imply that: rather it is a simple human presence on the sea (Figure 5.1). Nonetheless it is worth recalling that a part of the literature speaks only of sea uses with reference both to the uses *sensu stricto* and activities. This is justified by the fact that every kind of social behaviour towards the sea implies the exploitation of marine resources: naviga-tion also does the same because it exploits the sea as a *relational space*. The same can be said for cables and pipelines. Analysis that will develop from now on will be consistent with the latter approach: it will speak *tout court* of sea uses.

From the above definition it emerges that the sea use structure is a combina-tion of four sets: (i) the set of uses, (ii) the set of relationships between uses, (iii) the set of physical, chemical and biological elements, (iv) the set of relationships between sea uses, on the one hand, and natural elements, on the other. In order to investigate this complicated structure, the first phase consists of taking the *sea use framework* into account. According to the "global marine interaction model" [Couper ed., 1983, 208], the set of sea uses can be clustered at two levels: (i) cate-gories of sea use, (ii) individual sea uses. The result of this framework consists of 9 categories and 47 individual uses.

This taxonomic approach provides a general framework for the investigation of specific marine areas, both coastal and oceanic. For instance, it was applied to the Mediterranean Sea with the aim of disaggregating it in the marine areas [Vallega, 1988]. The framework of categories and individual sea uses are similar to those provided by the "global marine interaction model", as can be seen here-under.

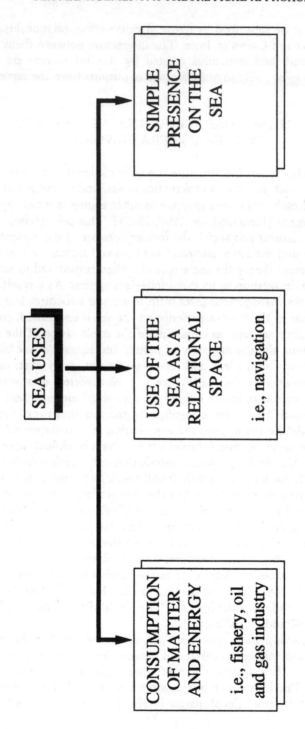

Figure 5.1 *The width of the concept of sea use.*

TABLE 5.1
The global marine interaction model

sea uses	
categories	*individual uses*
navigation and communication	navigation aids harbour/port shipping separation lanes cables
mineral end energy resources	sand and gravel dredging maintenance dredging separation drilling production platform coastal oil installations oil transportation pipelines ocean mining tidal energy
biological resources	demersal fishing pelagic fishing seaweed gathering aquaculture
waste disposal and pollution	solid waste nuclear waste incineration industrial effluent sewage oil pollution
strategy and defence	nuclear test zone firing/bombing ranges torpedo ranges submarine exercise areas minefields
research	fishery research marine geology oceanography archaeology
recreation	swimming/diving fishing yacht racing/cruising sailing ocean cruises
management: conservation	reserves marine parks
management environment	sea surface water quality ecology fish stocks sea bed subsea minerals wrecks

TABLE 5.2
The Mediterranean marine interaction model

sea uses	
categories	individual uses
navigation and communication	shipping
	separation lanes
	navigation aids
	ports
	offshore port terminals
	cables
mineral resources	sand and gravel dredging
	exploration drilling
	production plants up to 200 m isobath
	production plants beyond 200 m isobath
	coastal oil plants
	pipelines
	ocean mining
biological resources	demersal fishing
	pelagic fishing
	aquaculture
waste disposal and pollution	riverine discharge
	industrial outfalls
	urban sewage
	oil pollution
defence	exercise areas
research	archaeology
	scientific research
recreation	fishing
	yacht racing
	sailing
	marine cruising
protection	reserves
	marine parks

In order to make progress with this approach, first efforts to implement taxonomic methodology should be made and, secondly, a more detailed use framework should be sketched. As far as classification is concerned, it seems convenient to adopt a three level framework:

(i) *upper level*, categories of uses;
(ii) *intermediate level*, sub-categories of uses;
(iii) *lower level*, individual uses.

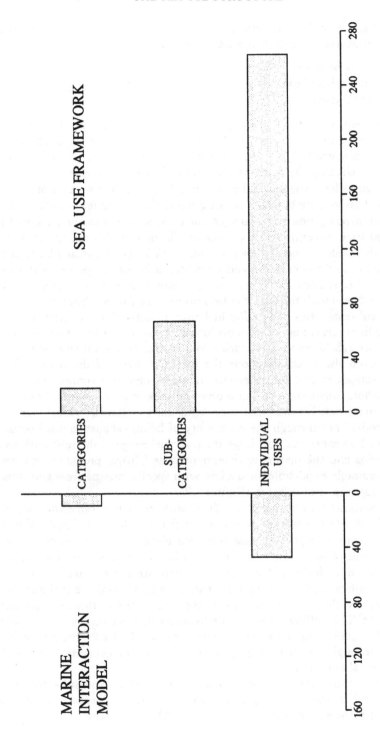

Figure 5.2 *The marine interaction model and sea use framework.*

On this basis a preliminary survey of current uses of marine resources and marine activities leads to a draft framework including

18 categories of uses;
67 sub-categories of uses;
263 individual uses.

As can be seen, when the whole matrix is worked out at the three levels, a very complicated network arises. Two methodological remarks seem appropriate.

First, the framework of uses here taken into consideration, as well any other framework sketched by the same method, has only a tentative role, i.e. it is only a reference point to be borne in mind with the aim of analysing sea uses of a specific marine area. Between the build-up of the general sea use framework and its application to the investigation and management of specific marine areas, a methodological feed-back relationship exists: the application of this sea use framework tends towards the setting up of a sea use framework in specific areas which, in their turn, contribute to the adjustment and implementation of the general framework.

Secondly, this framework is necessarily concerned with a specific temporal phase: it could deal with the past, the present and the future. When the cluster of categories, sub-categories and individual uses is defined with reference to the short or medium term, the expected evolution of the sea use structure may be investigated through forecasting scenarios concerning man-sea interaction.

Moving from these considerations the two first levels of the sea use framework—i.e. categories and sub-categories of uses—can be assumed. The framework as a whole, including also the individual uses, can be seen in *Appendix A*. It is much more detailed than that supporting the mentioned "global marine interaction model" not through the propensity of being exhaustive but because of the need to (i) be conscious how large the potential range of the relationships between societies and the marine environment is and (ii) to provide a framework which is as suitable as possible for dealing with specific marine areas and thus for implementing empirical research (Figure 5.2).

Some taxonomic problems that are dealt with in such a framework are listed in *Appendix A*. Here it suffices to point out that this kind of classification presupposes that the concept of sea use has been clarified. It is commonly thought that the concept of sea use is clear, in fact it is not so: it is very difficult to say what a use is. Both facilities (such as man-made structures, installations, carriers), and activities (such as navigation, fishing, exploration) are included in the list of sea uses without regard to the fact that the nature of these components is very different. This difference loses significance when—as occurs in sea management—both facilities and activities are taken into account with the aim of listing any forms through which human presence on the sea generates relationships with the natural environment.

From this stand-point, the above mentioned categories and sub-categories of sea uses, as well as the taxonomic assessment included in *Appendix A*, have been identified with the aim of contributing to the understanding of:

TABLE 5.3
The sea use framework

categories	sub-categories
1. SEAPORTS	1.1 waterfront commercial structures
	1.2 offshore commercial structures
	1.3 dockyards
	1.4 passenger facilities
	1.5 naval facilities
	1.6 fishing facilities
	1.7 recreational facilities
2. SHIPPING, CARRIERS	2.1 bulk vessels
	2.2 general cargo vessels
	2.3 unitized cargo vessels
	2.4 heavy and large cargo vessels
	2.5 passenger vessels
	2.6 multipurpose vessels
3. SHIPPING, ROUTES	3.1 routes
	3.2 passages
	3.3 separation lanes
4. SHIPPING, NAVIGATION AIDS	4.1 buoy systems
	4.2 lighthouses
	4.3 hyperbolic systems
	4.4 satellite systems
	4.5 inertial systems
5. SEA PIPELINES	5.1 slurry pipelines
	5.2 liquid bulk pipelines
	5.3 gas pipelines
	5.4 water pipelines
	5.5 waste disposal pipelines
6. CABLES	6.1 electric power cables
	6.2 telephone cables
7. AIR TRANSPORTATION	7.1 airports
	7.2 others
8. BIOLOGICAL RESOURCES	8.1 fishing
	8.2 gathering
	8.3 farming
	8.4 extra food products
9. HYDROCARBONS	9.1 exploration
	9.2 exploitation
	9.3 storage
10. METALLIFEROUS RESOURCES	10.1 sand and gravel
	10.2 water column minerals
	10.3 seabed deposits
11. RENEWABLE ENERGY SOURCES	11.1 wind
	11.2 water properties
	11.3 water dynamics
	11.4 subsoil
12. DEFENCE	12.1 exercise areas
	12.2 nuclear test areas
	12.3 minefields
	12.4 explosive weapon areas

TABLE 5.3—*continued*
The sea use framework

categories	sub-categories
13. RECREATION	13.1 onshore and waterfront
	13.2 offshore
14. WATERFRONT MAN-MADE STRUCTURES	14.1 onshore and waterfront
	14.2 offshore
15. WASTE DISPOSAL	15.1 urban and industrial plants
	15.2 watercourses
	15.3 offshore oil and gas installations
	15.4 dumping
	15.5 navigation
16. RESEARCH	16.1 water column
	16.2 seabed and subsoil
	16.3 ecosystems
	16.4 external environment interaction
	16.5 special areas and particularly sensitive areas
	16.6 sea use management
17. ARCHAEOLOGY	17.1 onshore and waterfront
	17.2 offshore
18. ENVIRONMENTAL PROTECTION AND PRESERVATION	18.1 onshore and waterfront
	18.2 offshore

(i) whether and how the exploitation of non-renewable resources can be minimized;

(ii) whether and how the exploitation of renewable resources can be maximized;

(iii) to what extent environmental impact can be minimized in the implementation of marine resource exploitation.

The sea use framework embraces three kinds of uses, which can be found (i) only in the coastal area, (ii) only in the ocean area, (iii) both in coastal and ocean areas (Figure 5.3). As has already been pointed out, in this chapter the whole range of uses is considered with the objective of paving the way for specific analysis of coastal (Chapters 7 and 8) and ocean (Chapters 9 and 10) uses.

5.3 THE SEA USE-USE RELATIONSHIP MODEL

Using current methodological approaches—such as the "global marine interaction" [Couper ed., 1983, 208–209]—the sea use framework leads to the setting up of a matrix (Xi, Xj) in which the relationships among uses are represented. This matrix can be based on either categories, sub-categories, or individual uses.

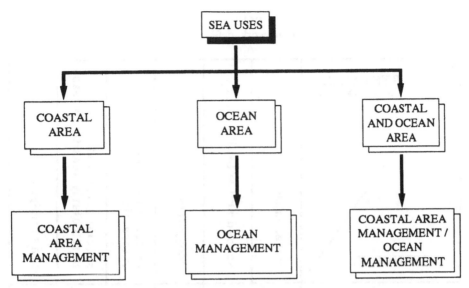

Figure 5.3 *Relationships between the nature of sea uses and main management categories.*

As a result, it identifies ranges of sea uses of various sizes according to the classification criteria adopted for setting up the sea use framework. The more sea uses are clustered the more detailed the analysis of their reciprocal relationships can be. The sea use framework, which has just been discussed, enables us to set up a detailed set of relationships among sea uses through a large matrix that may be called a *sea use-use relationship model*. There is no space here to display the model to its full extent and, *inter alia*, it does not seem necessary. As an example, the part of the model showing the relationships between "defence" (category no. 12 of the sea use framework) and "research" (category no. 16), which is represented in Table 5.4 with only the sub-categories (intermediate level of analysis), is the starting point for the investigation of relationships between these uses.

Subject 5.1
From chaotic states to order

"Current work in nonlinear dynamics finds order even in chaos: it develops an 'encyclopedia of bifurcations' which shows that seemingly chaotic states have their inner logic. Chaotic systems can be steered through interventions at critical points. Even more important from the viewpoint of understanding evolutionary processes is the empirical finding that, within the sweep of large-scale evolutionary processes, the outcome of bifurcations, though traversing chaotic states, is not entirely random. The statistical average is biased toward the

TABLE 5.4.
Sea use framework: defence and research

RESEARCH FIELDS X_j ⟋ DEFENCE X_i	water mass	seabed and subsoil	ecosystem	external environment interaction	special areas and particularly sensitive areas	sea management
exercise areas						
nuclear test areas						
minefields						
explosive weapons						

creation of structures that store more energy for a longer time, maximizing free energy and minimizing entropy. Without such a bias evolution would be a random drift between more and less organized states, instead of a generally one-way build-up of order and complexity through alternating phases of order and disorder, determinacy and indeterminacy". [Laszlo, 1988, 121].

This sea use-use relationship model is of course a product of a structuralist approach aiming at identifying sea use structures and investigating the relationships among the elements of the structures. Its objective is only to give impetus to sea management and, by its nature, it has two methodological constraints: (i) in itself the model is not a tool for dynamic analysis so, when the evolution of the sea use framework is to be investigated, additional or alternative methodologies are necessary; (ii) only use-to-use bilateral relationships can be represented, so

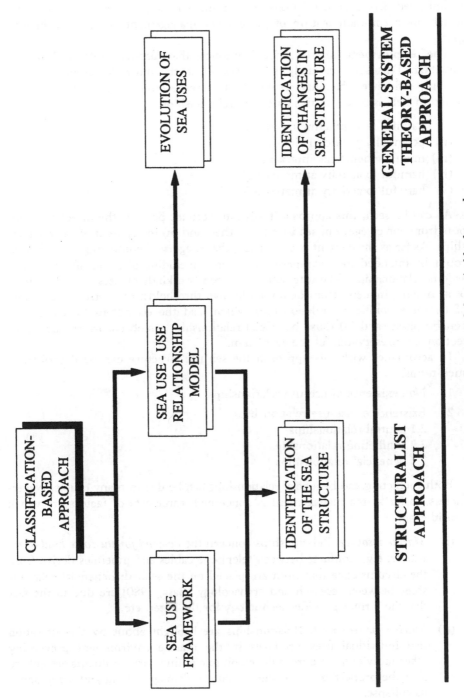

Figure 5.4 *Methodological phases and theoretical background*

complicated networks of relationships simultaneously involving a set of uses—which are a common feature of present sea management—cannot be investigated.

The employment of the model implies that use-use relationships are classified. This stage of analysis appears to be the most demanding since it strongly influences the final results. In this respect the "global marine interaction model" is based on this cluster of relationships:

(i)   harmful or conflicting interaction;
(ii) potentially hazardous interaction;
(iii) mutually beneficial interaction;
(iv) harmful to activity at matrix right;
(v) harmful to activity at matrix left.

As can be seen, this approach takes into account both (i) the direction of inputs from one category of sea uses to another, and (ii) the content of the relationships. As far as the content is concerned, the category of conflicting relationships could be intended in a comprehensive sense including also harmful and hazardous relationships. The interaction between two kinds of uses can be thought of as moving in a spectrum delimited by two thresholds, respectively consisting of (i) those conflicting relationships which lead the ecosystem to a morphogenetic phase, and (ii) those beneficial relationships which ensure the best protection and preservation of the ecosystem.

In accordance with this approach the sea use structure can be described in such terms:

1.   Non existence of use-use relationships

2.   Existence of use-use relationships
    2.1 neutral relationships
    2.2 conflicting relationships
    2.3 beneficial relationships

Both conflicting and beneficial relationships can be due to many causes. Nevertheless, as far as sea management is concerned, three sets of factors are worth considering.

(i)   *Marine resource.* Relationships concern the *demand for the same kind of resources*: e.g., conflicts between telephone cables and pipelines derive from the circumstance that both are placed on the seabed; beneficial relationships between research and archaeology [Bass, 1980] are due to the fact that the former provides technology for the latter, etc.

(ii) *Marine environment.* Relationships are brought about by the alteration that individual uses produces in the marine environment preventing other individual uses from developing: e.g., industrial effluents are able to alter the properties of the water column, damage fishing and bring about its collapse.

(iii) *Human context of the marine environment.* Each coastal and ocean area is characterized by specific social features which to some extent influence sea uses and the relationships between uses and the ecosystem. Political tensions, military activities [Westing, 1978; Morris, 1982], piracy [Birnie, 1987], terrorism [Pyeatt Menefee, 1988], risks of various kinds due to the existence of choke points [Alexander and Morgan, 1988], and other similar aspects should be regarded as the main features of this social context.

(iv) *Marine landscape.* Conflicts concerned with the marine landscape depend on the aesthetic values of the marine environment which people wish to maintain or create for developing a given use: e.g., the landscape generated by industrial uses of the sea, such as the loading and unloading of minerals through offshore terminals, is regarded as inappropriate for developing recreational facilities.

As an example, the relationships between research and defence could be as represented in the matrix displayed in Table 5.5.

5.4 THE SEA USE-ENVIRONMENT RELATIONSHIP MODEL

According to the 1982 Convention, the relationships between human presence and activities at sea should be managed with the main aim of protecting and preserving ecosystems and their physical environments [Boczek, 1986]. Regarding the relationships between marine resource exploitation and environmental management [Couper, 1978] as of primary importance, Smith and Lalwani [1984] suggested framing the relationships between uses and the environment by taking into account these components:

(i)   sea surface;
(ii) water column;
(iii) seabed;
(iv) coastal zone;
(v)   ecosystem.

As can be seen, three components—surface, water column and seabed—pertain to physical elements with which the ecosystem interacts. So these components, plus the ecosystem, constitute the natural context. The inclusion of the coastal zone in this framework brings forward a very interesting issue; when reasoning is focused on the environment—i.e. on the ecosystem and its physical elements—the coastal zone is regarded as the land-sea interface where (i) physical marine processes, such as the erosion cycle, generate effects and (ii) biological webs develop. By way of example, this approach leads us to conceive of the coastal zone as consisting of the inshore, foreshore and offshore [Camhis and Coccossis, 1982].

TABLE 5.5
Sea use-use relationships: defence and research

RESEARCH FIELDS X_j DEFENCE X_i	water mass	seabed and subsoil	ecosystem	external environment interaction	special areas and particularly sensitive areas	sea management
exercise areas	B2	A	B2	A	B2	A
nuclear test areas	B2	B2	B2	B2	B2	A
minefields	B1	B2	B1	A	B2	A
explosive weapons	B2	B2	B2	A	B2	A

A:　no existence of relationships

B:　existence of relationships

　　B1:　neutral relationships

　　B2:　conflicting relationships

　　B3:　reciprocally beneficial relationships

　　B4:　relationships beneficial to use x_i

　　B5:　relationships beneficial to use x_j

Subject 5.2
Defining pollution

"Defining pollution in an uncircular and workable manner can be difficult. Perhaps it is best to consider pollution as simply one type of environmental stress—the introduction by man of materials or energy into a system which causes unexpected or unpredictable change to take place, relative to what we consider the normal course of events. A legal definition of pollution, one that is acceptable to countries concerned with direct effects of pollutants on the health and well-being of its human residents, is necessarily anthropocentric." [Thacher and Meith-Avcin, 1978, 294].

In addition to these remarks the opportunity of considering the physical context as consisting also of the subsoil and distinguishing upper, intermediate and lower layers of the water column, are worth stressing. In conclusion, as a preliminary approach, the marine environment may be imagined as consisting of these components.

TABLE 5.6
The components of the marine environment

coastal	*ocean*
Land: permanently emerged land surface periodically emerged land surface	
sea surface	*sea surface*
water column: upper layers intermediate layers lower layers *seabed*	*water column:* upper layers intermediate layers lower layers *seabed*
subsoil	*subsoil*

As a consequence, each use should be evaluated according to this pattern with the object of realizing whether it is endowed with ocean and/or the coastal ecosystem and its physical elements.

CHAPTER 6

MANAGING THE SEA USE STRUCTURE

6.1 THE ROLE OF THE EXTERNAL ENVIRONMENT

Both ocean and coastal ecosystems evolve because they undergo natural processes acting on various scales. As a result, ecosystems should be viewed in an evolutionary context, within a "global change" framework [Vallega, 1990/a]. In order to assess the evolution of coastal and ocean ecosystems, attention may be drawn, *inter alia*, to three processes: (i) plate tectonic dynamics; (ii) climatic change and related hydrological cycles; (iii) sedimentary and erosion cycles. These have already been dealt with in Chapter 2, and only a few additional points are considered here.

On historical time scales plate tectonic theory as such is not useful, but the implications of plate dynamics for sea management are important. First, the presence of mineral resources on seabeds and subsoils depends on the geological stage which the coastal and ocean areas have reached. Secondly, plate movement, including convergence, divergence and subduction, is the main cause of seismicity and volcanism, which turn has a wide ranging implications, such as the presence and diffusion of evaporite rocks and risks of natural disasters.

The importance ascribed to the hydrological cycle has depended on the growing consciousness of the closed links existing between the evolution of this cycle and that of climate, and including both sea level rise and the extent of ice cover, which produce constraints on sea management (Figure 6.1).

The sedimentary cycle influences the evolution of the entire continental margin, in which the largest range of marine resources is located and where most human activities have been developed. In particular, the evolution of the geomorphological features of the continental margin has a profound influence upon urban and extra-urban waterfronts. In the context of management these processes—like every process acting on planetary or sub-planetary scales—may be considered as belonging to the external environment with which the sea use structure interacts, thus focusing upon the external environment rather than the structure. In accordance with general system theory, this is of great importance both in the evolution of the sea use structure and in establishing appropriate management measures. It is also worth emphasizing that the external environment has both natural and social components. Thus the evolution of the sea use structure with regard to the external environment can only be fully understood in terms of both natural and social processes, which take place in that environment. This increases the complexity of sea management.

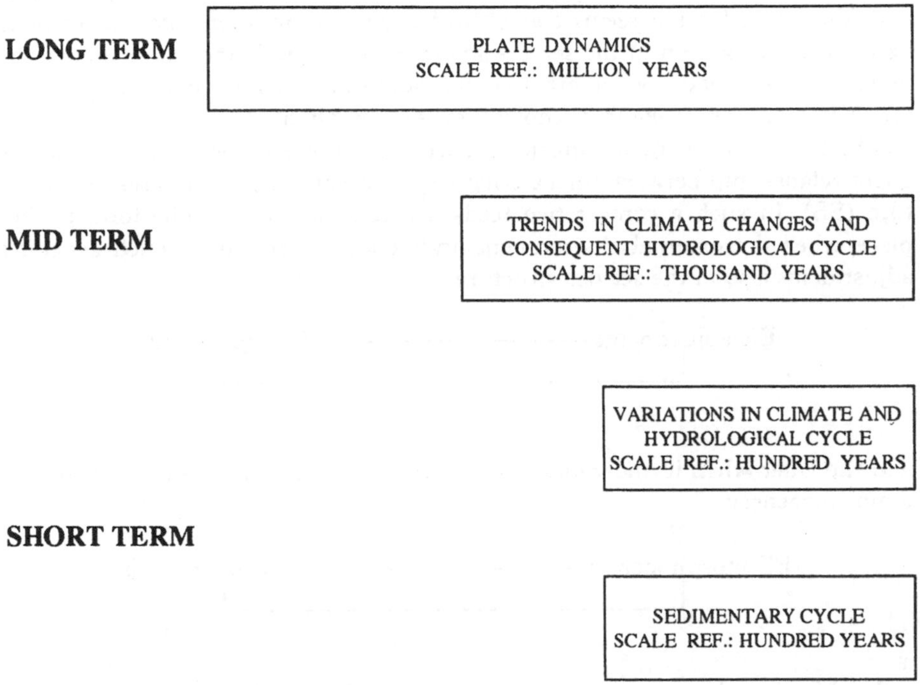

LONG TERM

> PLATE DYNAMICS
> SCALE REF.: MILLION YEARS

MID TERM

> TRENDS IN CLIMATE CHANGES AND
> CONSEQUENT HYDROLOGICAL CYCLE
> SCALE REF.: THOUSAND YEARS

> VARIATIONS IN CLIMATE AND
> HYDROLOGICAL CYCLE
> SCALE REF.: HUNDRED YEARS

SHORT TERM

> SEDIMENTARY CYCLE
> SCALE REF.: HUNDRED YEARS

Figure 6.1 *Reference time scales of the main cycles* involving the sea use structure.

The *natural external environment* is affected by the planetary processes which involve the sea as a whole. The climate is of particular importance since it influences hydrological and sedimentary cycles and determines the temperature of the upper layers of the water column. As a result, it is the main cause of the evolution of marine ecosystems and related food chains. The priority that, in general, the International Geosphere-Biosphere Programme (IGBP) gives to the interactions between climate and the ecosystem becomes especially significant in a maritime context.

The *social external environment* has four components of great importance: (i) *legal frameworks*, which since the early 1970s have been fundamentally developing to produce an unusually wide range of national claims; (ii) *technologies*, which are evolving in such a way as to create equipments capable of working in deeper and deeper seabeds and providing more diverse means for the protection and preservation of the marine environment; (iii) *economic and political strategies*, which consider the sea as having great potential; (iv) *culture*, which engenders explanations focusing on the sea as man's inheritance and a part of individual existential space.

Until now the literature has not extensively considered the external environment of the sea use structure and, *optimo iure*, has not provided concepts to deal with this issue. Here it seems enough to focus attention upon some interactions between the evolution of the sea use structure and that of the external environment. To this end two alternative matrices—between which, often unconsciously, managers make their choice—may be highlighted.

The first matrix is deterministic because it presumes the existence of a cause-effect relationship between the external environment (EE) and the sea use structure (SS). In such a context two feed-back relations occur. The former takes place when the external environment undergoes adjustments, which generates adjustments also in the sea use structure

$$\text{EE adjustments} \rule{4cm}{0.4pt} \text{SS adjustments}$$

and the latter when it undergoes morphogenesis, leading the sea use structure to a morphogenesis

$$\text{EE morphogenesis} \rule{4cm}{0.4pt} \text{SS morphogenesis}$$

In this view the sea use structure is conceived as a trivial machine: it makes only one response, and always the same response to the input from the external environment. Of course, the inputs from the sea use structure to the external environment are also regarded from the same viewpoint.

The second matrix is not deterministic because it does not presume a cause-effect relationship between the external environment and the sea use structure. *Apertis verbis*, on the one hand the sea use structure is also supposed to be able to make adjustments when the external environment profoundly changes. This could occur especially when the external environment changes at the legal, technological and/or cultural levels and the sea use structure does not react to these impulses.

$$\text{EE morphogenesis} \rule{4cm}{0.4pt} \text{SS adjustments}$$

On the other hand, the sea use structure is also supposed to enter a morphogenetic phase when it is subject to impulses from an external environment undergoing simple adjustments, because it is thought of as being capable of amplifying the effects that it receives from the external environment to the point of re-organizing itself. What has occurred in peripheral Japanese seas, particularly in bays, during the 1950s and 1960s is a significant example of such an amplification. At that time international trade and transportation was undergoing simple growth, namely, adjustments. Some large littoral cities—such as Tokyo, Yokohama and Kobe—reacted to these processes by creating artificial islands including port terminals, industries, and tertiary activities. As far as spatial assessments were concerned, that was a real morphogenesis since it introduced a new pattern for organising the waterfront.

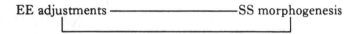

EE adjustments ————————————SS morphogenesis

In this view the sea use structure is conceived of as not being a trivial machine: it is able to give a range of responses to a single impulse from the external environment. Inputs moving from the sea use structure to the external environment may be approached in the same conceptual way.

Adjustments may consist of small changes in the set of elements and relations between elements while, when morphogenesis takes place, large changes in both sets occur, some elements disappear and new elements arise. Such adjustments guarantee the survival of the sea use structure, while morphogenesis leads to the birth of a new structure: according to the theory of macro-evolution this transformation is the final result of a bifurcation which the sea use structure has to face during its life. The more the involvement of the sea in human strategies and activities grows the more morphogeneses occur.

6.2 THE ORGANISATIONAL LEVELS OF THE SEA USE STRUCTURE

Both the growth of sea uses and the diffusion of complicated patterns of marine resource exploitation bring into play a new issue which will acquire more and more importance, namely, marine regionalisation. The concept of the region and, *optimo iure*, that of regionalisation are ambiguous *per se*, particularly when they are applied to the sea [Morgan, 1989*b*], but cannot be avoided. In the context of regional theory, regionalisation may be viewed as the product of the growing involvement of the sea in social activities and strategies, i.e. the product of an increasing complexity of the sea use structure.

The regional theory sustained by structuralist thought which developed in 1960s and 1970s has offered an interesting conceptual framework distinguishing single-feature regions, multiple-feature regions and compages [Whittlesey 1954]. From this initial approach the tendency to distinguish the real region

from the simple organized area has emerged. On the one hand, where only a partial involvement of marine resources occurs, there is a simple marine organised
area. On the other hand, where marine resources and the environment as a whole
are managed a real region occurs (Figure 6.2).

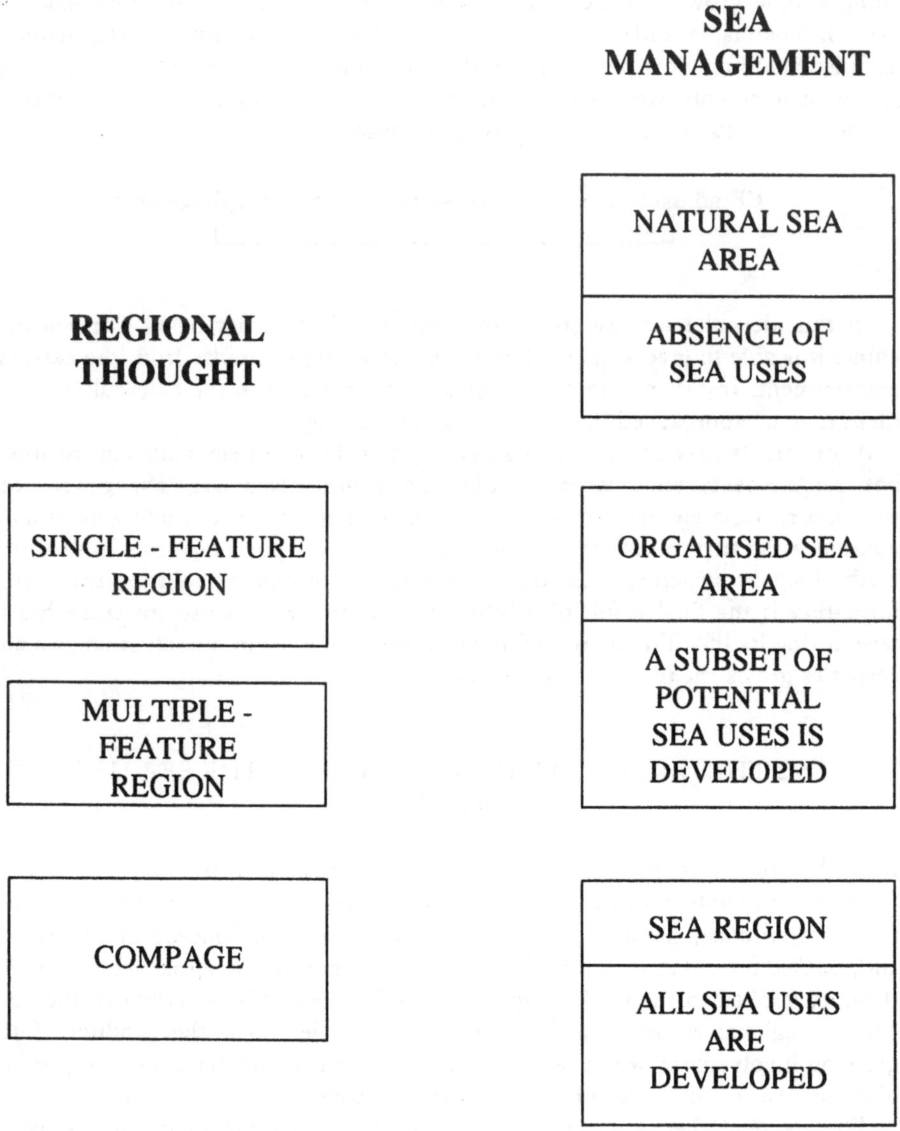

SEA MANAGEMENT

REGIONAL THOUGHT

NATURAL SEA AREA
ABSENCE OF SEA USES

SINGLE - FEATURE REGION

ORGANISED SEA AREA
A SUBSET OF POTENTIAL SEA USES IS DEVELOPED

MULTIPLE - FEATURE REGION

COMPAGE

SEA REGION
ALL SEA USES ARE DEVELOPED

Figure 6.2 *Regional thought and sea management*: the conceptual categories of the former
and the possible subsequent categories of the latter.

Starting from this basis and supposing that the sea may be affected by X_i uses, where X_i ranges from X_1 to X_n, it can be stated that:

(i)   where the sea is not subject to human activities there is a *natural sea area*;

(ii) where sea uses do not tend to X_n *an organised sea area occurs*;

(iii) where sea uses tend to X_n there is a *sea region*.

If these statements are agreed, the sea region—which could also be called *sea compage* according to the latin-based terminology from Whittlesey—appears in its reality: it is a much less frequent product of sea management, the diffusion of which will depend on how fast sea management evolves. Furthermore it occurs much more frequently in coastal zones where littoral, island and archipelagic communities have developed cultures and technologies strongly oriented towards the sea, with management approaches characteristic of a region.

How may the organized sea area be distinguished from the sea region? Explicitly speaking, given the range $X_1...X_n$ of uses, the problem consists of identifying a point in it that is the watershed between the organised sea area and the sea region. At the present stage of empirical research about sea management and regional thought there are no objective criteria to identify such a watershed. It should be noted that there may either be ample scope for answering this question or it may be thought that the question has no response.

In the abstract, it may be supposed that in the range (X_1, X_n) there is a point *m* beyond which the organised sea area acquires the features of a real region. In other words, given the range

$$X_1, X_2......X \ m, \ X_{m+1}......X_{n-1}, X_n$$

it should be noted that there is a sub-range (X_1, X_m) which relates to organised sea areas and a sub-range (X_{m+1}, X_n) which relates to sea regions. In explicit terms, the organised sea area occurs where human activities on the sea involve elements ranging from X_1 to X_m and the marine region occurs where also elements ranging from X_{m+1} to X_n are also involved in management (Figure 6.3).

As far as the spatial process—i.e. the diffusion of management patterns in sea spaces—is concerned, it may be recalled that, according to structuralist geographical literature:

(i)   the setting up of clusters of organised sea areas brings about marine spatial differentiation;

(ii) the creation of sea regions leads to sea regionalisation;

(iii) the passage from the assessment based on the X_1-X_m sub-range of sea uses to that based on the (X_{m+1}, X_n) sub-range lends form to the passage from marine spatial differentiation to sea regionalisation.

Subject 6.1
The terminological issue

According to the literature which has rapidly grown in the context of the 1982 UNCLOS and subsequent researches and initiatives in sea management two terms have arisen and spread: *the coastal area* and *the ocean area*. The former term is commonly used, particularly in the United Nations *milieu*, to indicate areas subject to national jurisdictions and the latter to indicate extra-national areas. The organisational level of these areas is not implied in these concepts. Nevertheless, where this level is concerned, regional theory is useful and, as a result, *area* acquires the meaning of a natural context less involved in social implications—i.e. less involved in human activities and facilities—while *region* acquires the specific meaning of a space so deeply involved in social implications that a *spatial system* takes place involving both human communities and the environment. In conclusion, the term "area" has different meanings in the literature on management compared to that on regional theory.

At the methodological level, according to taxonomy [Spence and Taylor, 1970], both areas and regions can be identified and spatially delimited by two alternative procedures:

(i) the aggregation procedure, in which detailed investigations on sea uses and social behaviour leads to assembling micro-spaces, called "taxonomic units";

(ii) the disaggregation procedure, through which a large marine space is divided into minor spaces endowed with the peculiarities of a sea area or a sea region.

6.3 TOWARDS MANAGEMENT PATTERNS

At this point two reference bases arise—the coastal area and the ocean area—both being related to very different approaches to sea management. The coastal area has been the starting point of sea management and, as a consequence, is the preferred issue of most of the literature on ocean affairs. Criteria that have been used to distinguish the coastal from the ocean area will be discussed in the next chapter. Here it suffices to recall two circumstances (Figure 6.4):

(i) the most frequently employed criteria are based on natural features, such as the outer limit of the continental shelf, and/or legal bases, such as administrative boundaries;

(ii) the extent of coastal areas has been progressively moved seaward involving larger and larger marine zones.

In order to sketch a general framework it should be borne in mind that:

(i)   the sea may or may not be subject to social influences;

(ii) in the former case organised areas or regions occur according to the organisational levels which the sea undergoes;

(iii) in the latter case the sea consists only of natural areas, i.e. areas characterised by a given ecosystem and its physical context;

(iv) these features are present both in coastal zones and in ocean areas.

As a consequence, the following conceptual scheme would seem to be justified.

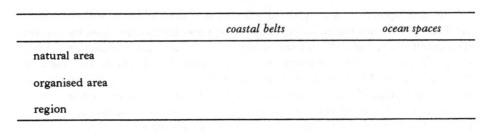

	coastal belts	ocean spaces
natural area		
organised area		
region		

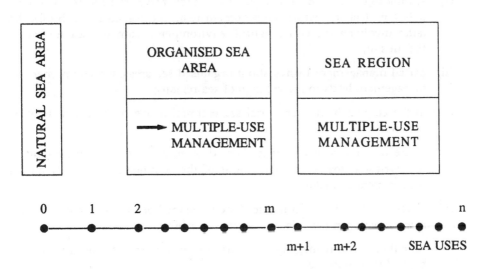

Figure 6.3 *Relationships between the number of sea uses and the level of sea management.*

In the present transition phase from the neo-industrial to the post-industrial stage (Chapter 3) there are certain trends:

(i)   more and more natural areas are involved in human activities and—as a result—they are being transformed into organised areas;

(ii) because of the growth of different types of sea uses and their diffusion, a growing number of organised areas are becoming real sea regions;

(iii) this process is diffusing especially in the coastal zones of both developed and developing countries;

(iv) ocean areas and regions are to be regarded as promising frontiers in the evolution of the involvement of the sea in economic and political strategies.

The development of an organised sea area into the ambitious and complicated product, which a sea region is, depends on the opportunities provided by the natural environment, national and corporate strategies and the jurisdictional framework. This leads to a re-consideration of the relationships between the natural and the legal aspects (Chapter 4). In that chapter the role of both the 200 nm line and of the outer edge of the continental margin were taken into account in order to assess how complicated the background against which sea management operates is. It is now appropriate to introduce sea management patterns into the reasoning.

To this end the statements supporting the identification of patterns can be considered thus:

(i)   sea management can be partial or comprehensive, the former involving only a part of the natural environment (e.g., the seabed and subsoil), the latter involving the entire natural environment, from the sea surface to the subsoil;

(ii) partial management brings about organised sea areas and comprehensive management leads to the creation of sea regions;

(iii) in the context of the territorial sea comprehensive management may be pursued;

(iv) where the continental shelf is claimed or agreed, from the outer limit of the territorial sea to the outer limit of the continental shelf only partial management can take place;

(v)   where the Exclusive Economic Zone is claimed or agreed, comprehensive management can take place up to its outer limit;

(vi) coastal or ocean management occurs respectively in the coastal zones and beyond their outer limits;

(vii) the separation line between the coastal and the ocean areas may be determined through natural or legal criteria, or through the combined application of both;

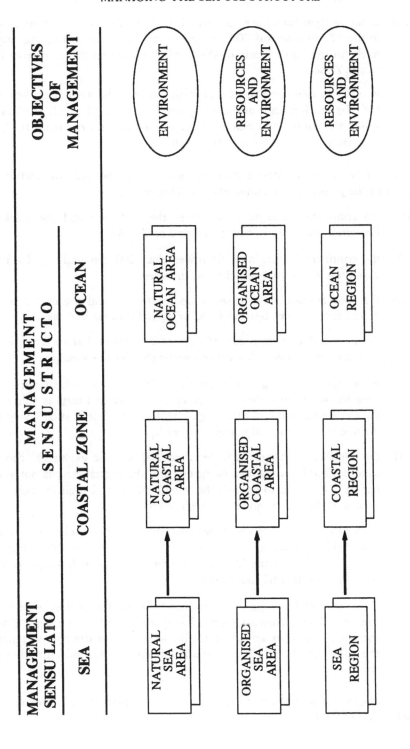

Figure 6.4 *The conceptual framework of sea management.*

(viii)   as far as legal limits are concerned, it is suitable to refer management to the 12 nm line and 200 nm line from the baselines relating respectively to the outer limits of the territorial sea and the Exclusive Economic Zone.

(ix)   in order to evaluate the relationships between the natural environment and the legal framework it is appropriate to distinguish the case in which the continental margin is narrower than 200 nm from the baselines from the case in which it is wider.

In accordance with the latest statement this interplay between the continental margin and the jurisdictional zones emerges (Figure 6.5):

(i)   the continental margin is narrower than 200 nm and the Exclusive Economic Zone has not been claimed or agreed;

(ii)   the continental margin is narrower than 200 nm and the Exclusive Economic Zone has been claimed or agreed;

(iii)   the continental margin extends up to 200 nm and the exclusive economic zone has not been claimed or agreed (border line case);

(iv)   the continental margin extends up to 200 nm and the exclusive economic zone has been claimed or agreed (border line case);

(v)   the continental margin is wider than 200 nm, the exclusive economic zone has not been claimed or agreed and the coastal state is able to opt for a criterion to delimit the (legal) continental shelf in such a way as to include the continental margin within it;

(vi)   the continental margin is wider than 200 nm, the exclusive economic zone has not been claimed or agreed and the coastal state is not able to opt for a criterion to delimit the (legal) continental shelf in such a way as to include the continental margin within it;

(vii)   the continental margin is wider than 200 nm, the exclusive economic zone has been claimed or agreed and the coastal state is able to opt for a criterion to delimit the (legal) continental shelf in such a way as to include the continental margin within it;

(viii) the continental margin is wider than 200 nm, the exclusive economic zone has been claimed or agreed and the coastal state cannot claim a continental shelf extended up to the outer edge of the rise, so a part of the continental margin is excluded from the national jurisdictional zone.

The management patterns coming from these combinations are concisely displayed in Appendix B.

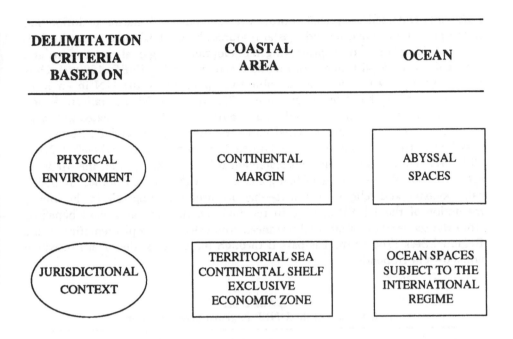

DELIMITATION CRITERIA BASED ON	COASTAL AREA	OCEAN
PHYSICAL ENVIRONMENT	CONTINENTAL MARGIN	ABYSSAL SPACES
JURISDICTIONAL CONTEXT	TERRITORIAL SEA CONTINENTAL SHELF EXCLUSIVE ECONOMIC ZONE	OCEAN SPACES SUBJECT TO THE INTERNATIONAL REGIME

Figure 6.5 *The basic criteria for the delimitation of the coastal area and the ocean.*

6.4 THE PROGRAMMES FOR REGIONAL SEAS

The analysis developed thus far is worth relating to the United Nations initiatives focused upon the management of regional seas (Figure 6.6). As UNEP [1982, 1, 37] stresses, the Regional Seas Programme put down its roots a long time ago, when in 1902 the International Council for the Exploration of the Sea (ICES) started making investigations in certain marine areas, especially in the North Atlantic and in the Baltic Sea: "while in most cases early regional activities began as research programmes, they frequently became a foundation for the management of marine living resources and, approaching the 1970s, for the control of marine pollution" [UNEP, 1982, 37]. These contributions were revealed as important when, in 1972, the United Nations Conference on the Human Environment held in Stockholm adopted guide-lines linking together environmental assessment, environmental management and supporting measures. In that context the need for regional approaches was self-evident. The UNEP, which was asked to carry on the UN action for environmental protection, chose ocean and coastal areas as one of its priority fields. This subject area was dealt with in three programmes relating to: (i) the Global Marine Environment, (ii) Regional Seas and (iii) Living Marine Resources. "The Regional Seas Programme was initiated by UNEP in 1974. Since then the Governing Council of UNEP has repeatedly endorsed a regional approach to the control of marine pollution and to the protection of marine and coastal resources by establishing a network of regional action plans and by promoting the adoption of regional conventions and protocols as the legal framework for these action plans" [UNEP, 1990, i]. How this programme was related to the objectives pursued by the UN in environmental protection has been clear since 1975, when the Mediterranean Action Plan was adopted. It demonstrated "that the basic concepts formulated at Stockholm can effectively foster regional co-operation among interested States and may benefit from the support of the United Nations system as a whole" [UNEP, 1982, 37]. To fully appreciate this statement it is worth noting that the Mediterranean Action Plan—in spite of being concerned with a space troubled by political, economic and religions conflicts—has acquired a leading role in the implementation of the UNEP activity in regional actions and has always benefited from the cooperation of all Mediterranean countries. The implementation of the Regional Seas Programme has passed through these stages, the details of which are shown in Appendix C.

TABLE 6.1
Action Plans of the UNEP Regional Seas Programme

Action Plan Conventions	adopted	come into force	goals (*)
Mediterranean	1976	1978	P, D
Kuwait	1978	1979	P
West and Central Africa	1981	1984	P, D
South-East Pacific	1981	1986	P
Red Sea and Gulf of Aden	1982	1985	P
Caribbean	1983	1986	P, D
East Africa	1985	—	P, D
South Pacific	1986	—	P

(*) Goals: P - environmental protection; D - development.
     From UNEP, *Status of regional agreements negotiated in the framework of the Regional Regional Seas Programme*, Rev. 2, Nairobi, 1990.

The Mediterranean Action Plan—in the context of which the Mediterranean Pollution, Blue Plan and Priority Actions Programme were worked out—was the initial phase for the setting up of a wide spectrum of plans involving a part of the Persian Gulf, the Red Sea and the Gulf of Aden as well as wide areas in the Atlantic and Pacific [Thacher, 1983; Hulm, 1983].

    This programme, which included many investigations and a stimulating set of experiences, can be considered from three points of view: (i) the nature of marine spaces to which it was related, (ii) the goals pursued in single marine areas, (iii) the relationships between UNEP's regional and global actions.

As far as the nature of areas benefiting from the Action Plans is concerned, it is self-evident that UNEP has been focused upon two categories of sea spaces: the semi-enclosed seas and some parts of the oceans having special relevance to the environmental implications of uses, such as land-based pollution sources and navigation. To a large extent a semi-enclosed sea consists of a homogeneous natural structure because its features are different from those of surrounding marine spaces. On the other hand, it occurs less frequently that parts of an ocean, such as the South Western and the South Eastern Pacific, have natural features so homogeneous that they can be distinguished from other surrounding ocean areas. The United Nations has established regional plans for some ocean areas— and, as a consequence, has defined them as regional—for the simple reason that the countries facing them believed it convenient to co-operate for the protection and preservation of the marine environment and, in some areas, also for the promotion of co-ordinated management patterns. In other words, the regional context has been identified through two variables: (i) the extent of coastal and ocean areas involved in the same range of environmental implications from uses; (ii) the extent of the coastal zones pertaining to the states interested in co-operating for environmental protection. The delimitation of the area involved by the Action Plan has been goal-oriented and influenced by the political context.

This circumstance would have only a theoretical value if it did not influence the management patterns. In fact, when the outer edge of the continental margin is considered as the potential seaward limit of the coastal area, it follows that some regional seas are to be considered as potential grounds for developing only coastal management because their beds only consist of the continental margin or the continental shelf. For instance, this occurs in the Red Sea and the Persian Gulf. On the other hand, where the regional sea consists of both the continental margin and ocean areas there is room to develop both coastal and deep-sea management. This occurs in almost every other marine space included in the UNEP programme.

These considerations allow us to shift attention to the objectives of the Action Plans. The main objective is the protection of coastal and ocean areas from pollution and in some cases—such as in Pacific areas—from contamination. This objective responds to the guide-lines of international co-operation formulated in Stockholm (1972) and tends to react—through regional conventions and subsequent protocols relating to specific range of pollution sources—to the environmental implications brought about by economic development during the post-war decades. Nevertheless, some conventions and protocols push UNEP's action towards the preservation of special areas, e.g. in favour of areas provided with special fauna and flora or an ecosystem which calls for appropriate measures. This has been done for preserving special Mediterranean, Central and South American, and North East African coastal areas (Appendix C). In these cases the spectrum of regional co-operation is basically more extended than that occurring where only the goal of combating pollution is the core of the action. Finally, the teleology of UNEP's action is much more extended where the Action Plan is also management-oriented, in the sense that they set up cooperation also with the aim of creating a common basis for coastal organization, e.g. implementing research on economic and social assessments and trying to promote the adoption of common criteria for developing coastal uses. This has taken place in the Mediterranean context, where the Blue Plan has led to the establishment of a common data bank, has encouraged investigations aiming at the setting up of scenarios concerning coastal regions and has become, as a final result, a background for the establishment and implementation of common criteria in the management of coastal uses. Where this enlargement of UNEP's action occurs, an intriguing concern comes to the fore, namely, the relationships between environmental protection and resource development. From this stand-point it is undoubtedly true that some of UNEP's experiences could provide useful contributions for seeing how, in practice, the harmonisation of these two goals, which theoretically do not seem compatible, can take place and which objectives it is able to achieve.

The global view of concerns—such as those that arose in the Stockholm Conference (1972), those that were set up in the context of UNCLOS III and those supporting general conventions for combating pollution—were the necessary background for the UNEP Regional Seas Programme and, *sensu lato*, for giving impetus to the regional approach to regional co-operation. On the other hand, regional conventions and their frameworks of co-operation, which were established by states involved in regional co-operation, have provided useful contributions towards the implementation of general guide-lines and global action for the global ocean environmental assessment. As a final result, "regional programmes and legal agreements that they foster may provide an important middle step between global principles and national implementation of those principles. Regional co-operation, which often provides the most suitable framework for an exchange of information and experiences, assistance and training, and an established political perception of the benefits to be derived from concerted regional action, seems to permit more readily the translation of principles and objectives into concrete actions and commitments" [UNEP, 1982, 29].

CHAPTER 7

THE COASTAL USE STRUCTURE

7.1 INTRODUCTORY REMARKS

Moving from the concept of the sea use structure (Chapters 5 and 6) the specific features of the coastal area can be taken into account. A short preliminary comment about terminological implications provides a useful starting point. Until now the expression *coastal zone* has been in use in the literature. This is probably due to the terminology introduced through the 1972 Coastal Zone Management Act in the United States legislation [Armstrong and Ryner, 1981, Chapter 4; Mitchell, 1986], which has strongly influenced national policies and legislation on management. Recently the expression *coastal area* has also come into use more extensively. For example, Vallejo [1988] speaks of coastal areas particularly with a view to emphasizing the differences which are encountered in moving from Coastal Area Management (CAM) to Ocean Management (OM). For two reasons this recent approach seems more appropriate: (i) the word *zone* has, as its main meaning, that of "a belt of the earth" and, secondly, has that of "a portion of the earth", while *area* means "a portion of the earth" and currently is used also as synonymous with "region"; (ii) since the analysis of sea management is methodologically linked to regional theory the concepts of "area" and "region" seem more appropriate than that of "zone".

The history of coastal area management has not yet been exhaustively written but all authors who have dealt with this subject agree that it is short and well developed. Mitchell [1982, 265] states that "some farsighted observers had recommended comprehensive state coastal conservation over a quarter of a century earlier, but the first significant modern state legislative action directed at an explicit coastal management problem was undertaken in 1959. In that year Texas passed an open beaches law designed to confirm public ownership of, and unimpeded access to, beaches". As far as the evolution of CAM is concerned, Mitchell [1982, 265–272] and Vallejo [1988, 205–206] sketch similar models. Mitchell identifies two phases: (i) the emergence of modern coastal zone management policy (1960–72); (ii) the post-1972 period, which has been influenced by the US CZMA (1972). Vallejo distinguishes: (i) the 1970s, during which "there was general recognition of the importance of marine resources for the

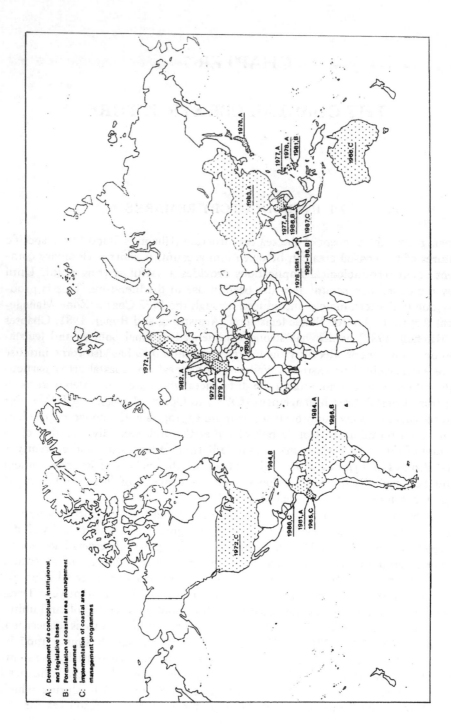

Figure 7.1 Development and diffusion of coastal area management programmes. Data from Vallejo, 1988, 208-209.

A: Development of a conceptual, institutional, and legislative base

B: Formulation of coastal area management programmes

C: Implementation of coastal area management programmes

economic growth of the states, an increase in scientific research, and for a sustained negotiation effort at the international level"; (ii) the 1980s, during which "there has been an incipient response from governments to the opportunities available, as well as a further recognition of the responsibilities involved in the newly acquired rights over Exclusive Economic Zones" (Figure 7.1).

These approaches provide starting points for constructing an historical model of coastal management. When the managerial system is viewed in relation to the theoretical and methodological background a three stage historical model can be sketched.

TABLE 7.1
Historical stage-based model of coastal management

stages	before 1970s	1970s	1980s and beyond
	pre-take off	take off	maturity
1. objectives	managing the shoreline	exploiting the coastal area	managing the coastal area
2. background (paradigm)	structuralist		from structuralism to general system theory
3. conceptual basis (coastal area regarded as...)	the shoreline	an administrative or geomorphological area	the continental margin and/or the 200 nm jurisdictional zone
4. approach	mono-disciplinary		pluridisciplinary
5. main scientific sectors involved	geomorphology, maritime transportation	... plus biology and law	.... plus ecology

This preliminary approach may be justified by the following reflections.

Pre-take off stage. Before the 1970s both the concept and practice of coastal management were concerned with a narrow zone which, according to Camhis and Coccossis [1982], consisted of the foreshore, inshore and offshore. The foreshore was conceived of as the zone extending between low and high tide marks in terms of average values per year. Both inshore and offshore were not conceived in a clear-cut way. Usually geomorphological criteria were adopted but they were considered *in a relative manner*: i.e. they were defined in such a way as to give impetus to the diffusion of settlements and facilities at the land-sea interface. Port structures and complementary structures—such as manufacturing plants processing unloaded raw materials—were the main elements. In particular, at that time offshore port structures under construction in Japanese bays (so called "artificial islands"), in the Persian Gulf (oil loading terminals) and in a few other parts of the coastal world were regarded as the most advanced elements of coastal management. Meanwhile in many parts of the world coastal management involved only inshore waters and the foreshore.

As a consequence, coastal management was mostly approached by geomorphologists and researchers on seaports and navigation. Since these disciplines are very distinct from each other from both theoretical and methodological points of view, investigation was mono-disciplinary and aimed towards managing specific categories of coastal activities and facilities, such as mercantile seaports and recreational activities. Analysis was structuralist, but in a reductive manner: attention remained focused on the functions of man-made structures, such as onshore and offshore port terminals, fisheries and navigation. The idea of the structure as the whole, consisting of coastal uses, the coastal ecosystem and its physical environment, was far removed from that approach.

Take off stage. During the 1970s many stimuli imposed an unexpected evolution on both the objectives and conceptual background of coastal management. In particular, the US Coastal Zone Management Act (1972) and the tendency to claim Exclusive Economic Zones—and exclusive fishery zones—exerted great influence. In the meantime offshore facilities—such as port terminals, oil and gas platforms, and multi-purpose artificial islands led to larger and larger marine zones being regarded as spaces integrated with littoral cities, seaports and industrial areas (Figure 7.2). As far as the objectives of sea management are concerned,

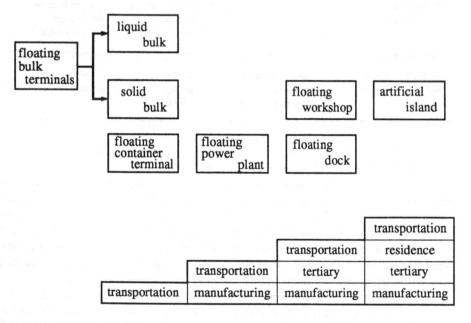

Figure 7.2 *The growth of coastal uses and the seaward advancement of coastal facilities.*

the implementation of sea uses was the main starting point and, as a consequence, one of the "major concerns was to protect the rather unique environmental position of the coastal area" [UN DIESA, 1982, 42]. The prospects of exploiting resources, which were encouraged and sustained by the establishment of national jurisdictional zones, had a leading role [Vallejo, 1988, 206]. In that context the approach continued to be mono-disciplinary, but a larger number of disciplines than in the past became interested in coastal management including biology, law and ecology. The conceptual basis remained structuralist. As for the spatial diffusion of sea management, at that time the conviction that developing countries would derive great advantages from marine resource exploitation were largely accepted giving impetus to sea management in developing areas [Levy, 1988].

Maturity stage. This stage started in the early 1980s and, at the present time, is still developing. The 1982 Convention, rapid technological advances, and the growing consciousness of the importance of coastal management for national policies have had important implications (Figure 7.3):

(i)   the objectives of coastal area management have changed, so that environmental protection and preservation have become objectives to be pursued together with the exploitation of marine resources and the implementation of sea uses;

(ii) coastal area management has evolved towards multiple-use patterns requiring the resolution of conflicts between uses [Archer and Knecht, 1987].

Meanwhile two criteria to delimit the extent of the coastal area have attracted more and more interest, both in the literature and in practice: (i) *a physical-biological criterion*, through which the continental (physical) shelf and, more specifically, the continental margin have been widely considered as *suitable contexts* for management, (ii) a *legal criterion*, which has regarded the 200 nm zone as the *marine milieu* for coastal facilities and activities. Hence, two consequences. First, the set of disciplines involved in coastal management has increased: climatology, sociology and urban planning also tackled issues related to some extent to this concern. Secondly, the scientific world has promoted multi- and interdisciplinary approaches. However, a structuralist basis is not appropriate for the creation of interdisciplinarity; hence the need to move towards general system theory approaches.

Bearing these historical features in mind coastal concerns can be dealt with using a general system-based approach congruent with the range of criteria and principles discussed in Chapters 1, 5 and 6. As a consequence, the following analysis will deal with: (i) the framework of coastal uses; (ii) the relationships between uses; (iii) the relationships between uses and the marine environment; (iv) the organisational levels of the coastal area.

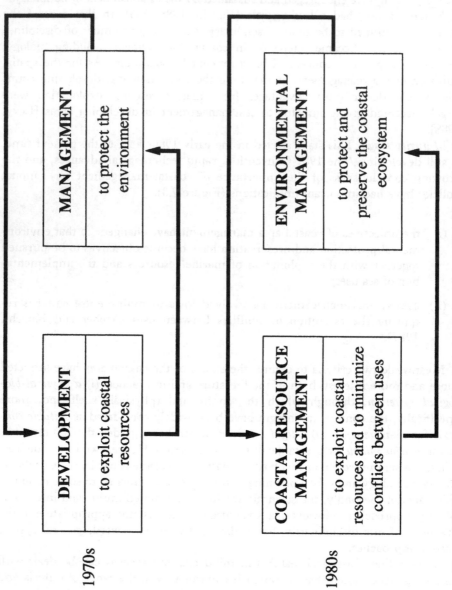

Figure 7.3 *The conceptual and terminological evolution of coastal area management in the context of the UN.*

7.2 THE COASTAL AREA: A TELEOLOGY-BASED VIEW

"The coastal area is the area of interaction between the land and the sea. It therefore includes terrestrial and marine resources, both renewable and non-renewable. It would thus include a landward component, the submerged lands and the continental shelf, and the superadjacent waters" [UN DIESA, 1982, 1 and 10]. This concept, which reflects the way in which the coastal area is commonly perceived, is both ambiguous and meaningful: *ambiguous*, since defining the coastal area as the land-sea interface is a tautology; *meaningful*, since it shows how the coastal area is viewed in relation to the combined exploitation of land and marine resources. As a consequence, in current thought on management the coastal area is not a thing existing *per se*—as an objective reality—but in accordance with the use of resources that littoral and archipelagic communities make according to social objectives. This implies that analysis begins by considering the goals of coastal management, namely, through a system-supported approach. In this context the coastal area should be thought of as a system, namely, *a coastal use structure changing because of* (i) *the impulses coming from its elements* and (ii) *the interaction with its external environment.* The coastal system has two components and two spatial extents (land and the sea).

The components are society and the ecosystem: links between them consist of chains of relationships moving from the former to the latter, and *vice versa.* The general concept of the bi-modular structure discussed in Chapter 1 acquires specific features in the coastal area. This reality is complicated, because often a littoral community is more or less closely tied to a plurality of ecosystems on land and in the sea.

In 1977, when coastal management was evolving rapidly, a report of the United Nations Secretary-General report [UN Document E/5971, 354] stated that this new economic frontier is the result of two main social objectives, *development* and *management* and, because of that, it is based on a hybrid concept, "combining elements of regional analysis and environmental management". This remark is appropriate since the complexity of coastal area analysis and management to a large extent depends on the need to deal with this issue through concepts and methodologies provided by regional science. This approach is multidisciplinary and interdisciplinary *per se*: insofar as it deals with coastal uses it is involved in social sciences; in the context of ecosystems of the coastal area and their physical environments it is concerned with natural sciences; insofar as it deals with the interactions between coastal uses and ecosystems it has to provide interdisciplinary methodologies, especially isomorphisms; to the extent that it considers objectives it should provide plans combining the need to use resources and also to protect and preserve ecosystems and their physical environments.

Development aims at the maximization of sea uses, while environmental management asks for the minimisation of the social impact on the ecosystem. The UN DIESA [1982, 4-8] rightly points out that the maximization of both economic efficiency and environmental quality are not meaningful since every economic objective produces an environmental impact. The question regards the extent to which coastal management should guarantee that fixed thresholds of environmental impact are not exceeded while expected economic objectives are reached. This is the basis of the *economic principle* according to which every coastal management pattern has a cost: "one cannot get something for nothing" [*ibid.*, 5]. As a consequence, society requires objectives for coastal management in relation to environmental impacts and has to take appropriate decisions. According to the *ecological principle* adopted in the United Nations analysis [1982, 17-8], the marine ecosystem is to be regarded as an organism in which all components are strictly tied to each other creating food chains and storing energy. The question is how to exploit both food chains and energy without over-stressing the ecosystem.

Subject 7.1
Stresses and changes in the coastal area

"One of the problems that frequently faces those involved in coastal management is that of distinguishing between natural and human-induced stresses and changes in the environment. No environment is static and natural systems are remarkably resilient to a broad range of stresses. Many natural systems are also in the process of making long-term adjustments to their environment. These adjustments may be in the form of ecological succession, a process by which plants and animals modify their environment, thus making it suitable to different species as the process advances; or, the changes may be in response to gradual long-term phenomena such as rising sea levels". [UN, DIESA, 1982, 8].

These statements mirror the state of the art in the late 1970s. In fact, during the 1980s two factors have become important: (i) societies have improved their perception of the relationships between their behaviour and marine ecosystems and have perceived more clearly than in the past the indispensability of protecting and preserving natural environments; (ii) technologies to manage the marine environment have progressed. Hence two consequences: (i) *ceteris paribus*, sea uses have decreased the capability of producing environmental stress; (ii) environmental protection-oriented engineering has increased its role to the point at which environmental management has been considered to be *an economically important objective*. This has strengthened the consciousness that coastal management acquires such different features that the relationships between the exploitation of coastal resources and the protection of coastal ecosystems should be studied on the dynamic level. This approach enables us to realize how ecosystems evolve in relation to social behaviour. In particular, it becomes clear that this interaction could also push the coastal ecosystem towards unexpected objectives implying undesirable adjustments and morphogeneses.

As the analysis of case studies advances, management objectives are investigated in different parts of the coastal world using theoretical and methodological frameworks. Recently dynamically-oriented analyses have emphasized profound change—a real bifurcation—in moving from conventional to new systems of goals: for example, in the investigation of Chinese coastal areas [Mukang H., Shusong Z. and Luiqing G., 1988]. On the other hand, "bifurcations" involving developing countries have been investigated by Mitchell [1982] and those occurring in developed countries have been set out in numerous researches. It is worth pointing out not only the width of the specific goals pursued in every coastal area but also the general framework resulting from them. Bearing in mind the United Nations analysis [1982] already mentioned and also taking into account the reasoning of Smith and Lalwani [1984, 232] the following framework seems to be justified—as a first approach:

1. *Development-oriented goals*
   1.1 Implementing conventional coastal uses;
   1.2 Setting up new coastal uses;
   1.3 Minimising conflicts between coastal uses;
   1.4 Making land planning secure;
   1.5 Making offshore facilities and activities secure.

2. *Environmental management*
   2.1 Protecting and preserving the marine ecosystem;
   2.2 Preserving special ecosystems and species.

7.3 THE DELIMITATION OF THE COASTAL AREA: A GOAL-BASED APPROACH

As is well known, criteria for delimiting coastal areas are numerous [Kenchington, 1990, 14-27]. The United Nations analysis [1982, 10-12] groups these into four sets:

i. physical criteria,
ii. administrative boundaries,
iii. arbitrary distances,
iv. selected environmental units;

and concludes: "it is clear that no single criterion is universally applicable, nor can one criterion meet all the requirements for an effective definition of the management area. Simplicity may be the virtue of using one criterion, while competitiveness and environmental significance may be the virtue of another definition". More detailed, but conceptually not far from this approach, is the framework (Table 7.2) formulated by Smith and Lalwani [1984, 236] grouping five sets of variables:

i.　meteorological and oceanographic factors;
ii.　geomorphological factors;
iii.　biogeographical factors;
iv.　uses;
v.　legal factors.

As far as physical criteria are concerned, the conventional approach takes one set, or a restricted number of these, into account according to the peculiarities of the area to be investigated. Researchers, particularly those interested in developing a system-based approach involving both natural and social environments, tried to overcome the employment of one or a few variables and to adopt criteria based on the consideration of processes. In this connection the methodologies set up by Drozdod [1990] are worthy of attention. He introduced the concept of

TABLE 7.2
Criteria for the delimitation of the coastal area

Criteria	Examples
1. meteorological/ oceanographic	limit of land breezes seawards limit of sea breezes landward salt/fresh water boundary in estuaries
2. geomorphological	water depths seawards tidal range limits of storm wave action landward seaward limits of breaking waves salt/freshwater boundary in estuaries
3. biogeographical	sea flooding limits seaweed limits seawards marine tolerant land plants salt marsh limits
4. uses	port limits dredging navigation channels fish farming installations waste disposal outfalls beach recreation
5. legal	internal waters territorial sea continental shelf

From Smith H.D. and Lalwani C., 1984, 236.

N.B.: This framework was formulated in the context of the analysis of coastal management in the North Sea. Considering also their coastal and marine spaces, the Exclusive Economic Zone should be included in group 5 (legal criteria).

the geosystem and focused analysis on the ecosystems of the land-sea interface taking into account these elements and related processes: "aerosol fall, rain and snow, chemical composition of plants and animals, soils, surface and ground waters as well as mud and other bottom sediments in coastal water bodies". According to these criteria, two categories of geosystems emerge pertaining respectively to (i) the coastal areas and (ii) the ocean areas. Special relevance is attributed to coastal geosystems adjacent to river mouths and those located between mouths.

At this point the delimitation of the coastal area can be viewed from two complementary standpoints: (i) from a coastal point of view, according to which the delimitation depends on the goals of coastal management; (ii) from a ocean point of view. The latter, discussed in the next chapter, will be profitably placed in relation to the former developed here.

In order to delimit the coastal area attention is centred on (i) the objectives of management; (ii) the opportunities provided by national jurisdictional zones; (iii) the opportunities and constraints provided by the natural environment. By way of example, as far as Smith and Lalwani's table is concerned (Table 7.2), the set 4 (uses) is related to the objectives, while the sets 1, 2, and 3 relate to opportunities and constraints. Because of the interaction between goals, opportunities and constraints, a range of achievable coastal management patterns arise: coastal communities make choices between them, looking towards, the maximizing of advantages and the minimizing of constraints. In fact, the coastal community develops management and, as a consequence, chooses between practicable management patterns according to the way in which it perceives the coastal area. Perception, although it has not still benefited from much attention in the literature, has great importance in coastal management. Where the coastal area is perceived as a *natural environment*, primarily to be protected and preserved, criteria rooted in the physical environment and the ecosystem are regarded as pre-eminent and the continental margin, or the continental (physical) shelf, are the main reference structures. On the other hand, where the coastal area is perceived in an *exploitation context*, national jurisdictional zones are the primary level delimitation criteria, and the territorial sea or the Exclusive Economic Zone are the main reference points.

As stated in Chapter 4, the solution is found where there is a jurisdictional zone covering the continental margin and enabling us to manage the sea as a whole—from the surface to the subsoil. The jurisdictional zone that most frequently ensures that is the Exclusive Economic Zone, the delimitation of which has acquired special importance [Attard, 1987]. As a consequence, where the Exclusive Economic Zone is claimed and covers the continental margin, coastal, inland or archipelagic areas benefit from the greatest number of degrees of freedom in establishing the goals of coastal area management. Incidentally, where the outer limit of the Exclusive Economic Zone extends beyond the outer edge of the continental margin the coastal state is able to exploit the mineral deposits of deep seabeds because they belong to its jurisdictional area. For example, this occurs in the Pacific side of the United States [Holtz, 1988].

7.4 THE COASTAL USE FRAMEWORK

Until now scientific thinking has not succeeded in describing a structure consisting of both society and nature. Thought based on general system-theory seems to be able to reach this fascinating goal but, at the present time, there is no evidence to justify stating that it is acquiring the capability of intensively investigating relations between physical and biological processes on the one hand and social processes on the other. As a consequence, it is not surprising that the current literature on the coastal area does not provide many analyses on the relationships between physical, biological and social processes and, *optimo iure*, does not explain the coastal system in *these terms*: attention is focused on sea uses, i.e. on social processes, placing the ecosystem and its physical environment is the background, or the opposite takes place. In fact, the coastal system is also too

TABLE 7.3
The coastal use framework

categories	sub-categories
1. SEAPORTS	1.1 waterfront commercialstructures
	1.2 offshore commercial structures
	1.3 dockyards
	1.4 passenger facilities
	1.5 naval facilities
	1.6 fishing facilities
	1.7 recreational facilities
2. SHIPPING, CARRIERS	2.1 bulk vessels
	2.2 general cargo vessels
	2.3 unitized cargo vessels
	2.4 heavy and large cargo vessels
	2.5 passenger vessels
	2.6 multipurpose vessels
3. SHIPPING, ROUTES	3.1 short-sea routes
	3.2 passages
	3.3 separation lanes
4. SHIPPING, NAVIGATION AIDS	4.1 buoy systems
	4.2 lighthouses
	4.3 hyperbolic systems
	4.4 satellite systems
	4.5 inertial systems
5. SEA PIPELINES	5.1 slurry pipelines
	5.2 liquid bulk pipelines
	5.3 gas pipelines
	5.4 water pipelines
	5.5 waste disposal pipelines
6. CABLES	6.1 electric power cables

TABLE 7.3—*continued*
The coastal use framework

categories	sub-categories
7. AIR TRANSPORTATION	7.1 airports 7.2 others
8. BIOLOGICAL RESOURCES	8.1 fishing 8.2 gathering 8.3 farming 8.4 extra food products
9. HYDROCARBONS	9.1 exploration 9.2 exploitation 9.3 storage
10. METALLIFEROUS RESOURCES	10.1 sand and gravel 10.2 water column minerals
11 RENEWABLE ENERGY SOURCES	11.1 wind 11.2 water properties 11.3 water dynamics 11.4 subsoil
12. DEFENCE	12.1 exercise areas 12.2 nuclear test areas 12.3 minefields 12.4 explosive weapon areas
13. RECREATION	13.1 onshore and waterfront 13.2 offshore
14. WATERFRONT MAN-MADE STRUCTURES	14.1 onshore and waterfront 14.2 offshore
15. WASTE DISPOSALS	15.1 urban and industrial plants 15.2 watercourses 15.3 offshore oil and gas installations 15.4 dumping 15.5 navigation
16. RESEARCH	16.1 water column 16.2 seabed and subsoil 16.3 ecosystems 16.4 external environment interaction 16.5 special areas and particularly sensitive areas 16.6 coastal management
17. ARCHAEOLOGY	17.1 onshore and waterfront 17.2 offshore
18. ENVIRONMENTAL PROTECTION AND PRESERVATION	18.1 onshore and waterfront 18.2 offshore

complicated to enable scientific thought to achieve adequate results. On the other hand, analyses of relationships *between* a given natural process and given features of social behaviour are frequent: e.g., the relationships between the coastal erosion processes and man-made structures, such as seaports, between the evolution of hydrological cycle and the management of estuarine facilities and between the sea level rise and land planning. What it is absent is the analysis of relationships as a whole.

As far as the upper component of the coastal system—i.e. coastal uses and the economic processes and strategies on which they depend—is concerned, research has set up the "global marine interaction model" which has been discussed in Chapter 5. Starting from that approach a *coastal use-use relationship model* can be sketched. It should be regarded only as a tentative product, the usefulness of which consists only of paving the way to developing reasoning on methodologies, especially on taxonomic procedures. With this approach it appears necessary, as a preliminary step, to group coastal uses. Starting from the general *sea use framework* discussed in Chapter 5, a set of *coastal uses* can be built. It is included in *Appendix D*.

As in the aforementioned general framework, the *coastal use framework* is based on three taxonomic levels: (i) categories, (ii) sub-categories, and (iii) individual uses. Only the uses tied to or oriented to the sea are regarded as coastal so the model does not embrace all the coastal uses located in the coastal area. Although it is larger than the similar frameworks up till now provided by the literature, it is only a partial and a tentative approach *vis-à-vis* the complexity of the sea use structure. In addition, it is to be regarded—more than the sea use framework is (Chapter 5)—only as a reference point for building models relating to specific marine areas. Here only the sets of categories and sub-categories are reported: the whole framework is displayed in *Appendix D*.

As has been stated with regard to sea uses (Chapter 5), the *coastal use framework* leads to the setting up of the *coastal use-use relationship model* (Figure 7.4). It consists of a square matrix (Xi, Xj) in which the set of relationships between uses are identified and grouped. Both the advantages and limits discussed in Chapter 5 are self-evident and very relevant in the coastal area: on the one hand, in recent history a very complicated network of relationships between coastal uses has arisen, particularly where human pressure has become very high; on the other hand, changes in many coastal areas are so profound as to require new assessments of coastal uses.

Subject 7.2
The objective of coastal area management

In the late 1970s, according to the United Nations Secretary-General, a manual for coastal area management should have included: a) fishing and mariculture; b) mining and minerals (including hydrocarbons); c) transport and shipping; d) foreign and domestic tourism; e) manufacturing; f) agriculture; g) forestry; h) settlement. "These activities enter directly into the flow of national income and product.

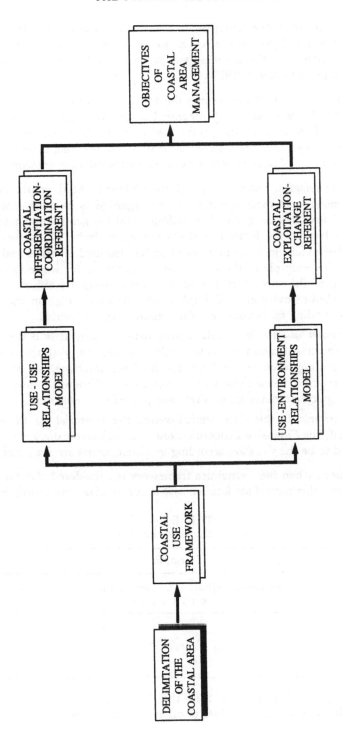

Figure 7.4 The phases of management-oriented coastal area investigation.

In addition to these activities, conservation, waste disposal and scientific research are also relevant to the coastal area, but do not directly enter into the development process." [UN Secretary General, Document E/5971, 1977].

For the objective of moving towards the analysis of use-use relationships the coastal use framework needs to be integrated with two matrices concerned with (i) the location of coastal uses and (ii) the environments they involve.

As far as the first variable (location of coastal uses) is concerned, three kinds of criteria are relevant for coastal management: geomorphological, environmental and legal

(i) *Geomorphological criteria.* Land, shoreline [Bird, 1985], continental shelf, slope and rise in their evolution can be regarded as its main components. Coastal uses can be grouped according to (a) the geomorphological zones in which they are located, and (b) the zones from which they are involved—which, in their turn, need to be classified [Fricker and Forbes, 1988]. An example of the latter case are pipelines for waste disposal: according to the ways in which deep current transports effluent, pipelines bring about environmental implications in places, e.g. on the slope or rise, extending far beyond the place where they are located.

(ii) *Environmental criteria.* In this environment the ecosystem is the reference point for classification procedures. Coastal uses are grouped according to whether they involve the land, the land-sea interface, sea surface, the water column, seabed and subsoil, since each of these components could host a specific ecosystem, or a relevant part of one ecosystem.

(iii) *Legal criteria.* In this view internal waters, the territorial sea, the continental shelf, the Exclusive Economic Zone and exclusive fishery zone are considered to be the variables according to which coastal areas are classified.

In conclusion, when the coastal use framework is considered also through the location of uses, this *coastal use location matrix* summarizes the approach.

TABLE 7.4
The coastal use location matrix

location		
geomorphological environmental c r i t e r i a		legal
coastal uses		

As an example of this approach the Table 7.5 shows the locational context of the exploitation and storage activities of oil and gas industry.

TABLE 7.5
Location of oil and gas plants

LOCATION ANALYSYS / COASTAL USES	geomorphological					environmental criteria						legal				
	land	shoreline	continental shelf	slope	rise	land	land-sea interface	sea surface	water column	seabed	subsoil	internal waters	territorial sea	continental shelf	exclusive economic zone	exclusive fishery zone
category 9: hydrocarbons																
9.2 exploitation																
9.2.1 extended onshore drills	●	●				●	●	●	●	●	●	●				
9.2.2 steel framed structures		●	●					●	●	●	●	●	●			
9.2.3 steel platforms			●					●	●	●	●		●	●	●	
9.2.4 concrete platforms			●					●	●	●	●		●	●	●	
9.2.5 hyperbuoyant leg platforms				●				●	●	●	●			●	●	
9.2.6 undersea manifolds					●					●	●			●	●	
9.3 storage																
9.3.1 waterfront facilities	●					●										
9.3.2 floating storages			●					●				●	●			
9.3.3 submerged storages			●							●		●	●			

At this point attention shifts to the environment upon which the coastal use acts. From the point of view of coastal management it is methodologically appropriate and practically relevant to take into account whether the coastal use involves the sea or the land. In this context four sets of uses can be distinguished:

(i)   *uses located onshore and not bringing about implications for the marine area*, such as industrial settlements not related with maritime transportation and not producing discharges;

(ii) *uses located onshore and having implications on the marine environment*, such as waste disposal [Leschine, 1988], seaports, recreational uses, sand and gravel dredging;

(iii) *uses located offshore and not involving the land*, such as navigation aids;

(iv) *uses practised on the sea and involving the land area*, such as marine transportation, oil and gas offshore exploration and exploitation [Harrison, 1983].

Uses causing interactions between land and the sea are worthy of attention since they particularly influence coastal management.

As a result, the analysis of use-use relationships may be made also taking into account the *matrix of spatial implications of coastal uses*, which may be conceived as follows:

TABLE 7.6
Spatial implications of coastal uses

	uses		
	land located	*sea located*	
		generating implications	
on land	*at sea*	*at sea*	*on land*
coastal uses			
....			
....			

As an example, Table 7.7. represents the framework of implications related to the sub-categories of the waste disposal. It is of course a tentative approach and, since it does not refer to individual uses, it also leads to the intermediate level of the classification of uses.

TABLE 7.7.
Spatial implications from waste disposal

COASTAL USES category 15: waste disposal	location			
	onshore		offshore	
	generating implication			
	on land	on the sea	on the sea	on land
subcategories 15.1 urban and industrial plants				
15.2 watercourses				
15.3 oil and gas offshore installations				
15.4 dumping				
15.5 navigation				

7.5 THE COASTAL USE-USE RELATIONSHIP MODEL

Analysis now moves from the coastal use framework to the relationships between coastal uses. Bearing in mind how this issue was developed in Chapter 5, where sea uses were considered in themselves, it is justified to group the use-use relationships in such terms:

1. Non existence of use-use relationships

2. Existence of use-use relationships
    2.1 neutral relationships;
    2.2 conflicting relationships;
    2.3 beneficial relationships.

For example, according to this approach the relationships between the installations for exploiting oil and gas, as well as storage and sea pipelines, on the
other, can be viewed as Table 7.8 shows.

TABLE 7.8
Relationships between oil and gas installations and sea pipelines

5. SEA PIPELINES x_j / 9. HYDROCARBONS x_i	5.1 slurry pipelines	5.2 liquid bulk pipelines	5.3 gas pipelines	5.4 water pipelines	5.5 waste disposal
9.2 exploitation					
9.2.1 extended onshore drills	**B1**	**B3**	**B3**	**B2**	**B2**
9.2.2 steel framed structures	**B1**	**B3**	**B3**	**B2**	**B2**
9.2.3 steel platforms	**A**	**B3**	**B3**	**B2**	**B2**
9.2.4 concrete platforms	**A**	**B3**	**B3**	**A**	**B2**
9.2.5 hyperbuoyant leg platforms	**A**	**B3**	**B3**	**A**	**B2**
9.3 storage 9.3.1 waterfront facilities	**A**	**B3**	**B3**	**B2**	**B2**
9.3.2 floating storages	**B1**	**B3**	**B3**	**B1**	**B1**
9.3.3 submerged storages	**B2**	**B3**	**B3**	**B2**	**B2**

A: no existence of relationships
B: existence of relationships
    B1:   neutral relationships
    B2:   conflicting relationships
    B3:   reciprocally beneficial relationships
    B4:   relationships beneficial to use x_i
    B5:   relationships beneficial to use x_j

In this framework conflicting relationships have an appreciable role. It is worth
considering that potential and actual incompatibilities between uses occur when:

(i) various uses are located in the same place: in this case one can speak of *locational incompatibility*; for example, conflicts between mercantile navigation and naval exercise areas;

(ii) to some extent various uses imply incompatible management patterns, bringing about *organisational incompatibility*; for example, conflicts between maritime transportation in oil and gas offshore areas (crude oil and oil product carriers, supply vessels, etc.), on the one hand, and yacht racing and cruising on the other;

(iii) a use provokes environmental impacts that another cannot tolerate producing *environmental incompatibility*; for example, the warm water discharges from coastal power stations changes the coastal ecosystem which do not make it possible to establish marine parks;

(iv) one use has a visual impact that the other use cannot tolerate (*aesthetic incompatibility*); e.g., this incompatibility occurs between coastal manufacturing plants and recreational facilities.

Also the *direction of relations* should be taken into account in order to realize from which use the input arises, bringing about a conflicting or beneficial relationship with another use. The literature places much weight on the directions: e.g., the "global marine interaction model" [Couper, ed., 1983, 208-9] considers the direction of two kinds of relations: harmful and beneficial. In fact, if one wants to deal with this issue on the general level, analysis should embrace three kinds of relationships between coastal uses, according to whether they move

(i)   from the use C_i to use C_j;
(ii) from the use C_j to the use C_i;
(iii) from the use C_i to use C_j, and *vice versa*,

where

C_i are the coastal uses included at matrix right, i being ranged from 1 to n;
C_j, are the coastal uses included at matrix left, j being ranged from 1 to n.

The case (i) implies that
$$C_j = f(C_i);$$

the case (ii) implies that
$$C_i = f(C_j)$$

and the case (iii) occurs when
$$C_j = f(C_i); C_i = f(C_j).$$

Cases (i) and (ii) prefigure simple patterns of relationships: they could be regarded as *series relationships*. On the other hand, case (iii) is too complicated to be referred to a given pattern of relationships.

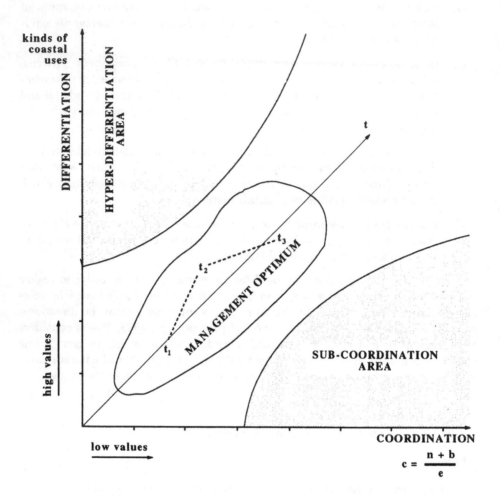

Figure 7.5 *The coastal differentiation-coordination model.*

At this point it is worth considering the evolution of coastal uses as a whole, i.e. the evolution of the social component of the coastal use structure. In order to facilitate reasoning on this issue the *differentiation-coordination model* can be taken into account. The stimulus to build this model has been offered by Le Moigne's approach to general system theory [1984, 249-250]. The role of this approach for the investigation of sea management as a whole has been discussed in Chapter 1: now attention should shift to the specific features which such a model could acquire to deal with coastal management.

As can be seen from Figure 7.5, the model consists of a triple-axis diagram: differentiation, co-ordination, time.

The *differentiation axis* (Y) represents the level of complexity that the set of uses acquires because of the growth of the exploitation of sea resources and environment. It is self-evident that differentiation depends on two variables: (i) the number of types of coastal uses; (ii) the technological level on which uses are managed. To make the model simple, the differentiation axis represents only the number of coastal uses. The values increase from the origin of the diagram onwards.

The *co-ordination axis* (X) shows the cohesive energy with which the coastal uses are provided. Conflicting relationships are regarded as reducing the co-ordination level and neutral and beneficial relations as acting in the opposite direction. As a first approach, co-ordination could be expressed by the relation

$$C = (n + b): c$$

where n is the number of neutral relationships; b is the number of beneficial and c the number of conflicting ones. The value of this ratio decreases moving from the origin of the diagram onwards.

The *time* axis represents the evolution of differentiation and co-ordination.

At a given time, which is characterized by a given level of technological advance, organizational patterns, etc.:

(i) the number of coastal uses cannot exceed a given threshold because beyond that the differentiation of the set of uses becomes too great to be managed;

(ii) the ratio (n+b):c cannot reduce below a given threshold because under that the co-ordination between coastal uses becomes too weak to enable us to manage the set of uses.

The set of values of differentiation placed above the differentiation threshold gives shape to the *hyper-differentiation area*. The set of values of the ratio (n+b):c, which are placed below the co-ordination threshold, gives shape to *sub-coordination area*. Between these two critical areas the *management optimum* exists. It is a management space within which the number of coastal uses and the relationships underlying them—and factors influencing them, as well—change without transferring the set of coastal uses towards critical points.

7.6 THE COASTAL USE-ENVIRONMENT RELATIONSHIP MODEL

Until now only one component of the coastal use structure has been considered, namely, society and its behaviour towards the coastal area. Now analysis should move to the other component, i.e. the coastal environment (the coastal ecosystem and its physical environment). From the point of view of coastal area management, analysis should not concern the coastal environment *per se* but the relationships between coastal uses and the environment: coastal ecology is to be considered strictly in relation to both the objectives, namely, organisational patterns and technologies of coastal management [Boaden and Seed, 1985].

As far as the classification of relationships is concerned, it is possible that groups similar to those pertaining to use-use relationships can be applied also to describe the *nexus* between uses and the coastal natural environment. As a consequence, neutral, conflicting and beneficial relationships can be introduced. Three vectors may be developed, consisting of:

(i)   inputs from natural to social environment;
(ii) inputs from social to natural environment;
(iii) inputs moving from the social to the natural environment, and *vice versa*.

When links are investigated at the dynamic level, series relationships are usually the main reference patterns and cumulative processes are identified, bringing forth growing damage or benefits to the ecosystem and/or the physical elements.

The components of the environment to which these kinds of relationships can be referred belong to the land and the sea. According to the approach developed in Chapter 5 (Table 5.6) this framework could be considered

TABLE 7.9
The components of the coastal environment

Land
permanently emerged land surface
periodically emerged land surface
sea surface
water column:
upper layers
intermediate layers
lower layers
seabed
subsoil

Relating the coastal uses to the components of the environment and taking into account the nature of relationships between these a logical matrix emerges, in which (i) lines embrace the components of the environment and (ii) columns include uses.

In the points of the intersections of lines and columns relationships are represented according to their directions, i.e.:

(i)   inputs from the natural environment to coastal uses;
(ii) inputs from coastal uses to the natural environment;
(iii) inputs moving in both directions.

For instance, the offshore oil and gas industry could bring about the framework of relationships represented in Table 7.10.

TABLE 7.10

Relationships between oil and gas industry and the coastal environment

9. HYDROCARBONS / COASTAL ENVIRONMENT		9.1 exploitation					9.2 exploitation					9.3 storage		
		9.1.1 jack-up rigs	9.1.2 semi-submersible platforms	9.1.3 drill ships	9.1.4 seabed wells	9.1.5 suspended wells	9.2.1 extended onshore drills	9.2.2 steel framed structures	9.2.3 steel platforms	9.2.4 concrete platforms	9.2.5 hyperbuoyant leg platforms	9.3.1 waterfront facilities	9.3.2 floating storages	9.3.3 submerged storages
land area	emerged	■					■					■		
land area	periodically emerged	■					■					■		
water column	upper layers	■	■	■	■	■	■	■	■	■	■		■	■
water column	intermediate layers	■	■	■	■	■	■	■	■	■	■		■	■
water column	lower layers	■	■	■	■	■	■	■	■	■	■		■	■
seabed		■	■	■	■	■	■	■	■	■	■		■	■
subsoil		●	●	●	●	●	●	●	●	●	●			

▲   inputs from the environment to coastal uses

■   inputs from coastal uses to the environment

●   inputs moving in both directions

The conviction that the main task of sea management is to keep under control environmental impacts and, *optimo iure*, to prevent collapses in ecosystems is spreading. This task is demanding in coastal area management because relationships taking place between society and the natural environment are very complicated. Keeping the environment under control is the usual task of management while preventing collapses calls for contingency plans—in the sense that the 1982 Convention (art 198) talks about—and, *lato sensu*, for plans capable of facing unexpected changes in the coastal area. That allows us to believe that, as a first approach, the inputs arising from coastal uses may be grouped into four categories:

— U_b, inputs beneficial to the coastal environment;

— U_n, inputs neutral for the coastal environment, because they do not bring about damages nor benefits to the ecosystem;

— U_r, inputs harmful to the coastal ecosystem and/or its physical context, hence the need for protection and preservation;

— U_z, inputs hazardous to the coastal ecosystem and/or its physical context, to which contingency and other similar plans could be referred.

In order to put coastal uses in relation to the coastal environment, the *coastal exploitation-change model* can be taken into consideration. It can be imagined as a three-axis Cartesian diagram in which:

— the axis Y represents the inputs arising from the coastal uses;

— the axis X represents the change in the coastal ecosystem and its physical environment;

— the axis t relates to time.

What is represented in axes Y and X can be made explicit by this approach.

Axis Y: the logic approach to the inputs from coastal uses can be based on the ratio:

$$X = \frac{(U_b + U_n) - (U_r + U_z)}{U_b + U_n + U_r + U_z}$$

Two ranges of values arise: *positive* when $(U_b + U_n)$ is higher than $(U_r + U_z)$; *negative* when $(U_b + U_n)$ is lower than $(U_r + U_z)$.

Axis X. When attention is centred on the marine environment the changes which it undergoes can be evaluated—as a preliminary approach, of course—in terms of BOD (Biological Oxygen Demand) and COD (Chemical Oxygen Demand).

Figure 7.6 provides a representation of the interactions between inputs from coastal uses and changes in the coastal marine environment. It may be stated that: (i) this diagram has only a conceptual value because it aims at encouraging discussion on use-environment dynamics; (ii) from this point of view it should be regarded only as a metaphor expressing the current ways by which the relationships between society and the coastal environment are examined; (iii) other parameters can be used to express the variables related both to uses and the environment unless the role of the diagram changes.

In this context the changes in the coastal ecosystem and its physical environment, which imply adjustments or morphogeneses, are regarded as being due to:

(i)   the increase of the ratio between harmful and/or hazardous coastal uses, on the one hand, and neutral and beneficial ones, on the other;

(ii) the increase of the global oxygen demand in the coastal sea.

That justifies the diagram including two environmental management areas (Figure 7.6).

(i)   *The adjustment area*. The set of neutral and beneficial coastal uses brings about inputs larger than those generated by the set of harmful and hazardous coastal uses. The latter set of uses not only is less extensive but is

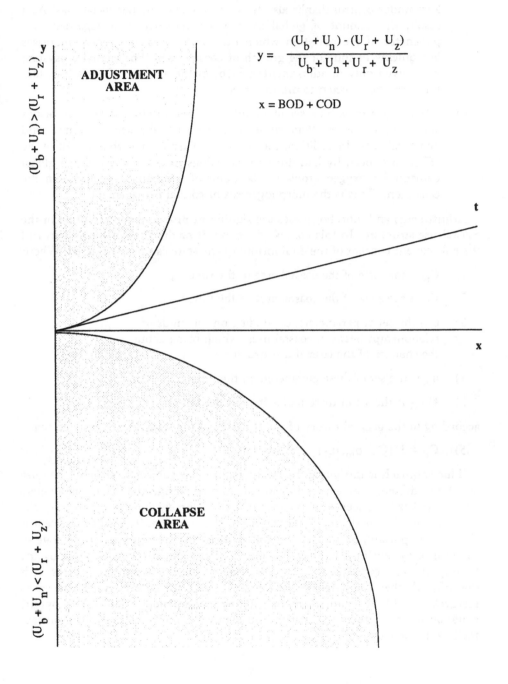

Figure 7.6 *The coastal exploitation-change model.*

kept under control thanks also to the role played by beneficial uses. As a result, the amount of global oxygen demand—which is regarded as a parameter of the pollution which the coastal area undergoes—increases less quickly in spite of the growth of coastal uses. The coastal ecosystem is involved only in adjustments which, in the absence of extra-uses factors, are not expected to produce risks.

(ii) *The collapse area.* The set of harmful and hazardous coastal uses brings about inputs larger than those generated by the set of neutral and beneficial uses. In addition, the latter set of uses is not able to reduce the effects produced by harmful and hazardous uses. As a result, the global demand for oxygen grows to the point of generating a collapse in the ecosystem. This is the morphogenesis of coastal management.

Adjustments and morphogeneses are significant phases of the evolution of the coastal use structure. In this context the concept of the state of the system and the subsequent concept of the equilibrium of the state can be considered. Where

1) C_{to} is the state of the coastal area in the time t_o;

2) C_{t1} is the state of the coastal area in the time t_1;

3) u is the set of parameters, depending on the kinds of relationships between coastal uses, which bring about the changes of the coastal use structure;

4) u_{to} is the set of these parameters in the time t_o,

5) 4) u_{t1} is the set of these parameters in the time t_1,

according to the general theory of the equation of the state, it can be stated that

5) $C_{t1} = F [C_{to}, u_{to}, u_{t1}]$

This approach is intriguing, but in its nature it is subject to criticism because of (i) the difficulty of expressing in quantitative terms the set of parameters which—as has been seen—by their nature can be largely expressed only in qualitative terms; (ii) the deterministic nature of the equation. The latter problem is the most important, of course. Reasoning based on the equation of the state of the coastal system implies that the time t_1 is thought of as determined by the time t_o and, as a consequence, that the future is the product of the present. It is doubtful whether this statement is compatible with the general-system theory so research should be cautious in developing methodologies supported by the equation of the state of the system and should search for other more satisfactorily approaches.

CHAPTER 8

MANAGING THE COASTAL AREA

8.1 THE MANAGEMENT SYSTEM

As the need to implement the rationale of coastal area management has been spread throughout the developed and developing worlds, particularly to give impetus to environmental protection and preservation [Levy, 1988; Kenchington, 1990, 60-74], *management systems* have been given increasing attention in the literature. In this context efforts have been made to relate decision-making systems to the goals of coastal management. To this end Mitchell [1982, 308–310] formulated a three dimensional model. The first axis represents the *policy structure variables*: management patterns range from "strong national and/or regional government control" to the exclusive role of private interest groups. The second axis refers to *administrative variables*: they range from bodies and agencies engaged in substantive problems—such as coastal erosion, waterfront development—to agencies and bodies "with broad functional responsibilities such as economic development, transportation, or land planning". The third axis pertains to *policy orientation*: in this context economic development and environmental protection and preservation are the extremes of the range. The result is that each management system has a role due to its place in the diagram, i.e. according to how much it is influenced by private or public administrative interests, deals more or less with specific concerns, and aims at implementing uses or protecting the environment.

According to Smith and Lalwani [1984, 237] the "management system" can be represented by a matrix consisting of two dimensions: (i) the level of organisations related to the spatial extent of planning and management; (ii) the use sectors, in which authorities operate. Thus international, national, regional and local authorities are identified bearing in mind that this framework varies from country to country. The use sectors include organisations concerned with seaports, maritime transportation, aquaculture, recreation, defence, etc., i.e. in coastal use sectors. In this context the need to establish the optimum scale in planning and management emerges as one of the most important issues [Jolliffe and Patman, 1985, 19–23].

Subject 8.1
The spatial scale

"A fundamental issue that confronts all coastal states concerns the spatial scale at which planning and management takes place. On the face of it, an holistic approach to coastal zone matters, with an emphasis on 'overview' as a pre-requisite to problem solving, should promote the optimization of resource use—working from the whole to the part. However, the apparent logic of this approach does not always appear to prevent decisions that stem from the part rather than the whole. This raises the question concerning whether or not there is an appropriate size for authorities with jurisdiction in the coastal zone. There are numerous factors that influence, or at least should influence, the answer to this question. Some problems or projects can be satisfactorily handled at a site-specific level, e.g. the disposal of sewage effluent; other problems can only be resolved at an international level, e.g. the offshore disposal of radioactive waste in the North Sea. In between these two extremes, one recognises an array of circumstances in which planning scale appears to become diffuse and arbitrary". [Jolliffe and Patman, 1985, 19–20].

The literature tends to investigate coastal management systems in terms of three categories of variables:

(i)   legal variables, i.e. the sets of *laws and regulations* which pertain to various levels (from the international to the local);

(ii) *authorities* considered through their spheres of operation;

(iii) subsequent assessments of *coastal users*.

As attention shifts from decision-making centres to their actions it seems appropriate to distinguish two operational levels, the former consisting of behaviour influenced by regulations, the latter of behaviour effectively displayed. Coastal area management patterns are more or less congruent with conventions and regulations, depending on the difference between law and practice. Thus there are varying degrees of capability in preventing and resolving conflicts between users: from this stand-point, the United States is an important national case [Armstrong and Ryner, 1981, Chapter 4]; the North Sea is an interesting international case [Uthoff, 1983]; while Canada consists of a significant example of bilateral involvement [Day and Gamble, 1990].

On the basis of the general system-inspired approach [Le Moigne, 1984, 129–149] the coastal management system may be regarded as consisting of three subsystems, decision, information and operations (Figure 8.1). Decision centres, which form the upper level of the management systems, have three functions:

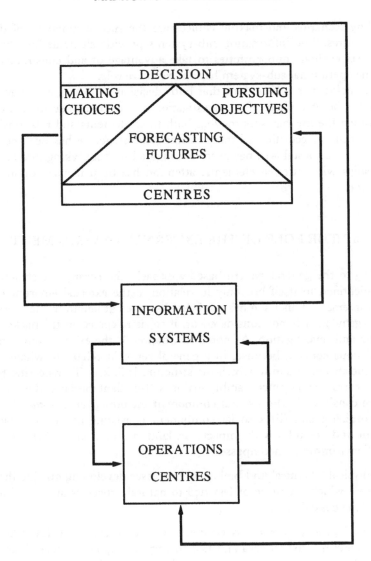

Figure 8.1 *Decision-makers' systems*, according to Le Moigne (1984, 129-49).

(i)   they *make choices* between alternative possibilities, i.e. alternative strate-
      gies for the use of the coastal area and the behaviour towards both the
      ecosystem and its physical environment;

(ii) they *foreshadow* the range of management goals which imply that alternative
      futures are envisaged and, in particular, that multiple-use management is
      regarded as the main reference basis [Burbridge, Dankers and Clark, 1989];

(iii) they *make strategies* founded on choices between alternative goals.

Building scenarios and making choices are the special functions of decision-making centres. The information sub-systems provide elements by which decision centres evaluate opportunities to take advantage of and constraints to deal with. The operational sub-system has an executive role.

At this point it is self-evident that both decision processes and management systems are the final phase of an investigation of the coastal use structure aiming at explaining the mechanisms moving both their elements and relationships between elements. According to this reasoning the structure has been considered in itself, i.e. the coastal area has been investigated without taking into account its relationships with external elements: attention has been focused upon endogenous factors.

8.2 THE ROLE OF THE EXTERNAL ENVIRONMENT

According to the general system-based approach, the coastal use structure cannot be identified in itself but only in relation to the external environment with which it interacts. This issue has not been discussed at length in the literature on sea management: it is not considered or, if so, is relegated in the background of reasoning and investigation. As noted in Chapter 1, the external environment is an ambiguous concept because, in a general sense, it relates to whatever exists outside the structure and to which the structure is linked. Transferring this concept to coastal management ambiguity is self-evident because the coastal use structure consists of: (i) a natural component including *terra firma*, the sea and the atmosphere: and (ii) a social component consisting of facilities, activities, behaviour and attitudes with impacts on land, sea and atmosphere. Hence the external environment encompasses:

(i)   physical, chemical and biological processes developing outside the coastal area, which are more or less tied to natural processes and social activities of the coastal area;

(ii)   social processes, such as economic and political strategies, taking place outside the coastal area and tied to varying degrees to coastal settlements and economic activities, physical, chemical and biological processes.

Relationships between the coastal area and its external environment pertain to at least three categories: (i) series relationships moving from the coastal area towards the external environment; (ii) series relationships moving from the external environment towards the coastal area; (iii) feed-back relationships.

From the theoretical point of view there is no great difficulty in identifying the external environment of the natural component of the coastal area: it consists of every external geomorphological, chemical and biological process and the factors to which this coastal component is related. As a preliminary approach, three sets of processes and elements can be taken into account.

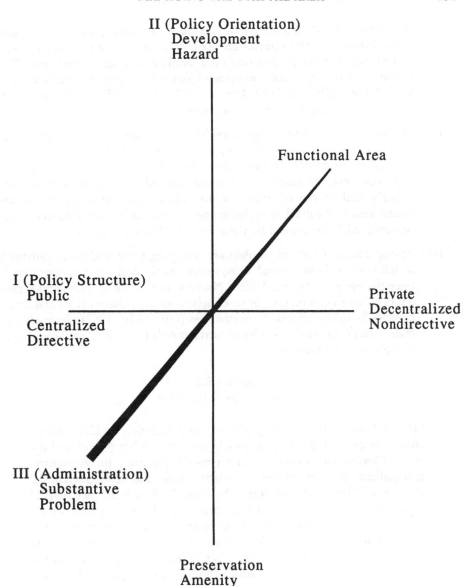

II (Policy Orientation)
Development
Hazard

Functional Area

I (Policy Structure)
Public

Centralized
Directive

Private
Decentralized
Nondirective

III (Administration)
Substantive
Problem

Preservation
Amenity

Figure 8.2 *Decision-making centres, policies and objectives of coastal area management.* From Mitchell, 1982, 309.

(i) As far as *global change*—regarded as world-wide transformations in the ecosystems that have being developed since about 18,000 years ago—is concerned, the major impulses from the external environment are due to climatic change. Both the rise in temperature, influencing sea-level rise, and changes in the chemical composition of the atmosphere and in regional and areal climatic conditions have had great influence in coastal areas. Meanwhile the interaction between climatic changes and the system of ocean currents has profound implications for coastal seas. The human response to these processes—particularly that to predicted sea-level changes [Orford, 1987; Carter, 1987]—is one of the most stimulating issues of present coastal management.

(ii) *Physical, chemical and biological cycles* are considered by the literature as reference bases in order to develop consciousness of the complexity of the external environment. In this context hydrological, sedimentary and erosion cycles may be assumed as the processes of the external environment strongly tied to the evolution of the coastal area. In other words, the coastal area is involved in cycles acting in more or less large spaces: it can be imagined as influenced by them and/or influencing them.

(iii) *Specific cycles of natural elements* are acquiring more and more capability of influencing both coastal ecosystems. As is well known, particularly since the early 1970s, the United Nations, as well its agencies and organisations, have been making great efforts to identify these cycles and investigate subsequent effects in specific seas [GESAMP, 1982]. *Inter alia*, attention has been paid to cycles of heavy metals (mercury, lead, cadmium), phosphorus and nitrogen.

Subject 8.2
Sea-level rise in the USA

"In the United States nearly 65% of the population in U.S. marine coastal states, or 102.5 million people, now live within 50 miles of the coast. Coastal communities must provide facilities, infrastructure, and policies to support the population base and maintain coastal amenities. This study concerns the manner in which state coastal policymakers and institutions have begun to address the issue of accelerated sea level rise. The current scientific uncertainties are compounded by the lack of uniformity and direction in the responses of federal agencies and policymakers. While the federal government has made a concerted effort to study the scientific aspects of climate change and sea level rise, it has not provided clear policy guidance or incentives to states or local governments regarding appropriate responses. The threat of sea level rise further complicates and exacerbates the process of planning in coastal communities". [Klarin and Hershman, 1990, 144].

The analysis of the role of chemical elements in geomorphological and biological contexts allows us to take social behaviour into consideration because of the pollution of the atmosphere and waters. At this point reasoning shifts to the social component of the external environment. This issue has far-reaching theoretical implications. Thus it is proper to bear in mind that the natural component of the external environment can be understood in an objective sense, because both processes and elements can be identified and measured through instruments. For instance, as far as the erosion cycle is concerned, one can identify the elements (waves, tides, currents, geological and geomorphological features, etc.), their action within the cycle and the spatial extent of the cycle. As a consequence, it is possible to identify what part of this extent is located outside of the coastal area.

On the contrary, when the social component of the coastal area—i.e. facilities and activities—is taken into consideration a different conceptual framework emerges. The external environment of a society does not exist *per se* but to the extent it is determined by the society itself. Thus each human community, explicitly or not—i.e. either in a regulated or random manner—establishes its goals and, as a consequence, it creates its own external environment. More explicitly, the external environment is *functional to the strategies of the leading groups and decision-making centres* of the community.

Both the natural and social components of the external environment are not necessarily spatially contiguous with the coastal area: e.g., the pollutants by which the coastal area is affected can be produced also from very distant areas, and the pollutants generated in the coastal area can affect distant regions. It happens more often that the social external environment of the coastal area consists of areas more or less distant from the coastal zone: e.g., with respect to the case of the exploitation of offshore oil and gas fields, as well as defence and navigation, the coastal area is linked with areas all over the world. The external environment, especially its social component, is dynamic and has an extent which is difficult to identify. In this context it needs scarcely be mentioned that, at the present time, a large number of coastal areas are evolving in such a way as to continuously re-create their external environment.

These theoretical results, to which the general system-based analysis leads, justify believing that efforts should be made in order to move coastal management patterns away from deterministic approaches and to leave more and more room—both in investigation and management—for the interaction between the coastal area and its external environment. This should aim at identifying or foreshadowing the goals to which this interaction could aim. What was discussed in abstract terms in Chapter 1 regarding the relationships between the sea use structure and its external environment is worth applying to the coastal area. As a preliminary approach, four main cases can be taken into consideration (Figure 8.3).

(i) *Stable external environment and stable coastal area.* This is the simplest case. The external environment continues to give the same impulses to the coastal area which, in its turn, does not change its outputs. As a consequence, the coastal system is stable.

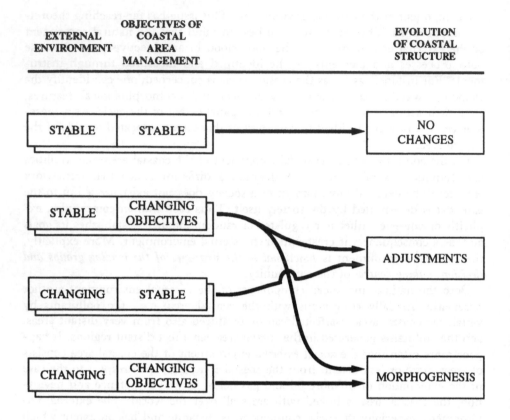

Figure 8.3 *Exogenous and endogenous factors of the coastal area evolution.*

(ii) *Stable external environment and changing goals of the coastal area.* New endogenous factors, which are brought about by changing strategies of decision-making centres and leading social groups, re-orient the coastal area, moving it towards new goals. According to the endogenous factors the coastal area undergoes simple adjustments or real morphogenesis, the latter being capable of converting the structure into a new one. The external environment may not be much influenced by the changing goals of the coastal area or, on the contrary, it can undergo adjustments. In the latter case feed-back relationships can develop between the coastal area and the external environment.

(iii) *Changing external environment and stable coastal area.* Because of new inputs from the external environment the coastal area is pushed only to changing some goals. If so, it undergoes only adjustments that, in their turn, generate new types of outputs. Feed-back relationships between the coastal area and its external environment can take place.

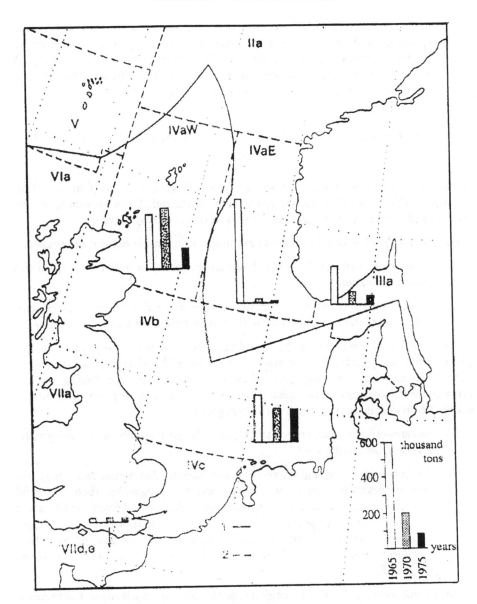

Figure 8.4 *The areal division of the North Sea according to herring catches*. From Coull, 1988, 118.

(iv) *Changing external environment and changing goals of the coastal area*. Since the 1970s this complicated case has been occurring in many coastal areas, both in developed and developing countries. The complication is due to the fact that, on the one hand, changes in the external environment bring about new influences in the coastal area and, on the other hand, because of endogenous factors the coastal area is pushed towards new goals. Endogenous and exogenous impulses interact generating an instability phase during which: (i) the coastal area undergoes morphogenesis; (ii) outputs from the coastal area change and, to some extent, influence the external environment; (iii) feed-back relationships probably take shape.

8.3 THE OBJECTIVES OF COASTAL MANAGEMENT: AREA AND REGION

At this point in the analysis, what was discussed in general terms (Chapter 5) on the organisational level of marine spaces can be related to coastal management. In a general sense, marine spaces are involved on three main levels:

(i)   they are not affected by human presence, so a simple *natural area* exists;

(ii) they can be affected by a small human presence as well activities on the sea, so an *organised area* occurs;

(iii) human activities can be developed to the point that they, together with their natural context, bring about a real system, giving form to a *region*.

As littoral organization and the exploitation of coastal resources advance, organisational complexity of these areas also grows and real regions appear. This process is occurring in many parts of both developed and developing worlds. Until now the concept of the coastal region has not been diffusely used in the literature on coastal management for two reasons:

(i)   researchers have not perceived the need to go through the organisational levels of the coastal area;

(ii) the concept of the region, and the subsequent distinction from the concept of organised area, come from regional geographic theory, a field upon which attention has not been intensively focused from the literature on coastal area management. In fact, when the literature on sea management—and *optimo iure* that on coastal management—speaks of "regions", it intends simply to speak of areas, whatever their organisational level (Figure 8.5).

As already noted (Chapter 6), regional geographic theory leads to distinguishing: (i) the single-feature region, identified through one variable (e.g., human settlements), (ii) the multiple-feature region, identified through a set of variables (e.g., land uses, settlements, traffic flows), and (iii) the compage, identified through all the organisational elements. Moving from this general approach to that specifically concerned with sea management it was stated that:

(i)   supposing that X_i are the elements, natural and human, of the sea organisation;

(ii) the organised sea area occurs where sea organisation involves elements ranged between X_i and X_m;

(iii) the sea region occurs where sea organisation involves also the elements ranged from X_{m+1} and X_n;

(iv) where X_m is a threshold established according to the main goals of sea management.

Focusing attention upon the coastal area it should be agreed that management is planned and developed with the aim of exploiting coastal resources without jeopardizing the coastal ecosystem and its physical context. As a result, the organisational level to which the coastal area is related—or expected to be related—is coastal resources: in explicit terms, physical, chemical and biological elements. In this context it is opportune to distinguish:

(i) the natural elements involved in coastal management;

(ii) the human elements, i.e. activities and facilities, through which coastal management take place.

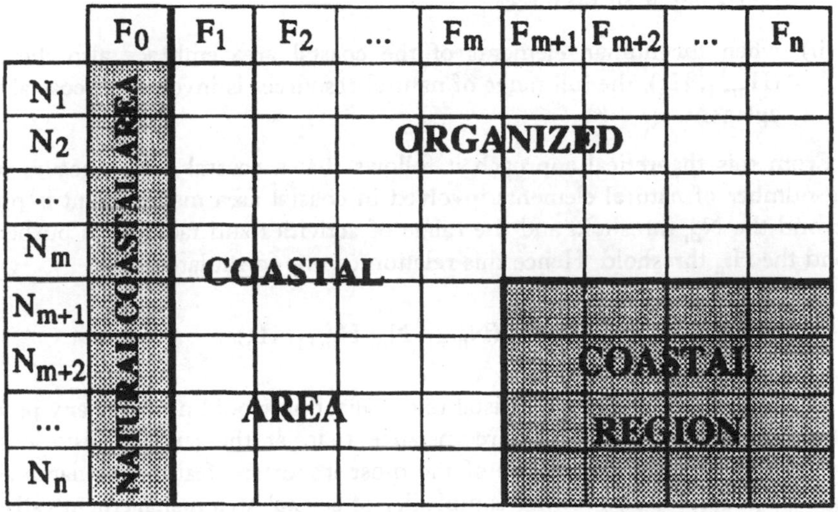

	F_0	F_1	F_2	...	F_m	F_{m+1}	F_{m+2}	...	F_n
N_1									
N_2				ORGANIZED					
...									
N_m		COASTAL							
N_{m+1}									
N_{m+2}							COASTAL		
...		AREA					REGION		
N_n									

Figure 8.5 *The complexity levels in coastal area management,* according to the growth of coastal uses. (N, natural elements; F, functions.)

Considering natural elements it is possible to know where there is a simple coastal organised area, or a coastal region exists—or where it is about to do so. In other words, the existence of an area or a region depends on the number and kind of natural elements involved in coastal management. It can be stated that:

(i)   supposing that N_i are the natural elements;

(ii)   a coastal organised area exists where the range (N_1, N_m) is involved in coastal management;

(iii) a coastal region exists where also the range (N_{m+1}, N_n) is involved.

Considering the human elements of coastal organisation it is possible to know how complex it is and, as a consequence, one can go through the levels of coastal organisation. The number of elements through which coastal organisation can be established and implemented depends on several variables and particularly on:

(i)   the level of available technologies;

(ii)   the features of the coastal ecosystem and its physical context;

(iii) the goals that the littoral community wants to achieve.

Supposing that human elements (H_i), through which coastal organisation *can be* set up, range from H_1 to H_n, it can be stated that *in a given historic stage,*

(i)   there exists a set of human elements, ranging from H_1 to H_m, that do not involve the coastal environment as a whole, i.e. they do not imply the exploitation of all resources;

(ii)   when the human elements of the coastal area embrace also the range (H_{m+1}, H_n), the full range of natural resources is involved in coastal management.

From this theoretical approach it follows that a coastal region occurs where the number of natural elements involved in coastal area management is pushed beyond the N_m threshold and the range of activities and facilities is pushed beyond the H_m threshold. Hence this relation can be formulated:

$$C = F(N_{m+1}, N_n; H_{m+1}, H_n)$$

Theoretical approaches to coastal regionalization would not have any practical interest were it not for the circumstance that, at the current historic stage, coastal area management is one of the most important features of national and regional policies. The growing complexity of coastal area management calls for a comprehensive approach and the consciousness that in many parts of the world a new and unexpected product, the coastal region, is spreading, leading to the development of *coastal regional theory*. It is enough to note the wide range of implications which arise from the establishment of legal frameworks in some regions—such as the East Asian seas [Prescott, 1987a]—to agree with this prediction. The evaluation of the environmental state of some major regional seas—as a comparative analysis shows [Clark, 1986, Chapter 10]—strengthens this statement.

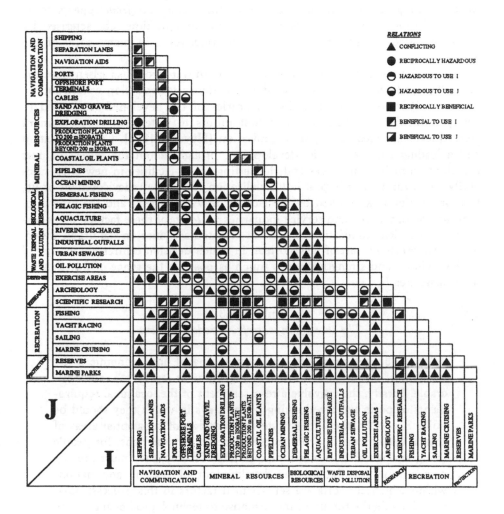

MEDITERRANEAN SEA

SEA USES
REFERENCE : 1985 ONWARDS

Figure 8.6 *The complexity levels of the management in semi-enclosed seas: the Mediterranean case.* From Vallega, 1990a, 198.

8.4 URBAN WATERFRONT: THE CORE OF THE COASTAL AREA

When a strong spatial organisation comes to light it is evidently provided with a core which assembles powerful and efficient decision centres and leading functions and gives impulses to the surrounding space thus keeping it under control. As far as networks of cities are concerned, this core consists of a *top level central place*, i.e. the city, or metropolis, endowed with the highest functions in the network. Shifting attention to the coastal area the *urban waterfront* appears as its *functional core* [Hoyle, 1988], especially at the present time, characterised by quick and strong transformation of urban waterfronts in many parts of the world.

As is well known, growing interest in the urban waterfront has arisen since the late 1960s and early 1970s because of the tendency to redevelop waterfronts in the United States, in other leading countries of the developed world and in newly industrialised countries. The first inputs to waterfront redevelopment were due to changes in seaport organisation [Hayuth, 1988]. Containerisation had a leading role because its development brought about the retreat of port functions from historic places and the setting up of facilities in new sites. Secondly, the evolution of littoral industrialization led to the redevelopment of littoral industrial areas marked by the lack of heavy industries and the growth of high technology-based and service industries. *A latere* of these factors, other impulses, which lead to location of non sea-oriented activities on the urban waterfront have created a sort of new horizon in which decision centres are interested in maintaining and strengthening as much as possible the range of functions in the coastal area.

Subject 8.3
The changes in waterfront redevelopment

"The challenges presented by the new spatial and economic order which port evolution and urban redevelopment represent are considerable. Firstly, they involve a re-assessment of the locational requirements of ports and a rethinking of the ways in which they should be built. Secondly, they necessitate the redesign of substantial areas of coastal settlements in order to create better environments for people living and working in maritime quarters. Thirdly, they must provide new stimuli for the further development of regions and nations within which cityports are set. Critical factors in this process, however, include the timing of the response to technological change, both in ports and cities; the broad ecological environment within which these changes are set; and the economic and political influences which underlie spatial and social change". [Hoyle, 1988, 3].

The urban waterfront, conventionally regarded as the interface between the seaport and the urban area, is involving larger and larger spaces. It may be expected that: (i) in many coastal areas, *in the short term* the planning of the urban waterfront will also involve urban spaces surrounding the port-city interface, and (ii) particularly because of population growth in coastal areas [Edwards,

1989], *in the mid and long term* will lead to the planning of the interface of whole metropolitan and megalopolitan areas [Stewart, 1970]. If this process takes place an unusual task will be faced by research, because both analysis and planning will be based on this conceptual matrix:

—   first coordinate: urban functions;

—   second coordinate: waterfront functions;

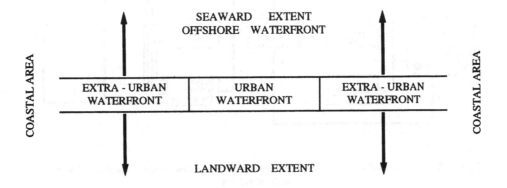

Figure 8.7 *The evolution of the waterfront in the context of the coastal area.*

implying that (i) the role of the waterfront will take form according to the urban and metropolitan assessment as a whole, and (ii) planning will acquire its role in the context of the evolution of the cityport within the urban area.

As attention shifts from the urban waterfront to the surrounding land-sea interface, the extra-urban waterfront emerges (Figure 8.7). It consists of urban settlements and structures of various kinds—from the littoral roads to the piers of recreational and fishing ports—which, since they are not so intense as in urban areas, generate less human pressure and environmental impacts than on the urban waterfront. In spite of that, at least two reasons converge into giving impetus to the evolution of the extra-urban waterfront.

(i)   On account of the world-wide growth of littoral populations and the progressive extent of urban areas, extra-urban areas are undergoing urbanisation and need planning as the present urban areas do.

(ii)   Because of growing welfare and subsequent changes in social behaviour, extra-urban areas are progressively affected by the setting up of recreational facilities, second homes etc. In this context human settlements and facilities cause impacts on the landscape and environment, which call for protection and preservation by means of an appropriate code of conduct.

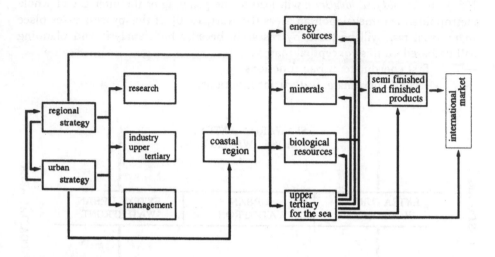

Figure 8.8 *Urban and regional strategies aiming at strengthening the waterfront and creating coastal regions.*

In conclusion, current reality is marked by some processes closely related each other: (i) the progressive extension of the urban waterfront; (ii) the growing complexity of its organization; (iii) the progressive involvement of the extra-urban waterfront in residential and non-residential facilities and human pressure; (iv) in many areas the conversion of the extra-urban waterfront into an urban one. Some relevant spatial implications derive from this.

The land part of the coastal area extends landward of the urban and extra-urban waterfronts. It encompasses (i) the spaces *actually* affected by the expansion of settlements and facilities and (ii) the spaces *potentially* affected according to the evolution which the coastal area undergoes. As far as coastal area management is concerned, greater interest attaches to the seaward extent of the coastal area which, in its turn, includes the *marine spaces actually and potentially involved*, the former including all offshore facilities and the current uses of the coastal sea, the latter including both facilities and uses which are expected to be set up. As a consequence, the goals of national policies and regional strategies (Figure 8.8) for waterfront development become the reference basis for building scenarios of the evolution of the core of the coastal area.

CHAPTER 9

THE OCEAN USE STRUCTURE

9.1 THE NATURE OF THE ISSUE

In approaching ocean management Vallejo's considerations [1988, 210-211] can be recalled: ocean management "has remained a theoretical concept discussed in some forums and by a few scholars who have anticipated the magnitude of the task involved in the formulation and implementation of an OM programme." Initial inputs can be found "in the *Pacem in Maribus* proceedings since 1970. In 1973, the Fabian pamphlet and a debate on sea-use planning in the UK parliament introduced, for the first time, the use of the term in a political context. Other countries, particularly The Netherlands, followed the idea and worked out a conceptual basis, producing a series of papers and publications that enriched the literature and filled a serious gap" [*ibid.*, 210-211]. Vallejo concludes that "current OM activities indicate that the take off from theory to practice has not yet occurred. There are, however, indications that we are reaching the beginning of a new phase, as indicated by strong technological, economic, social, and political patterns that presage changes, and with them the appearance of political, institutional, and planning decision-making directed to the effective incorporation of the EEZ within the framework of national development planning". [*ibid.*, 213].

As far as the history of ocean management is concerned, Vallejo stresses [*ibid.* 208-209] that the first "development of a conceptual, institutional and legislative base" is due to the 1969 Stratton Commission Report. As for policy the first step consisted of the 1972 US National Marine Sanctuary Program. As can be seen, this program was established in the same year in which the US Coastal Zone Management Act entered into force. Nevertheless these events are very different from each other in relation to the content of management: the Coastal Zone Management Act was the initial step in the creation of comprehensive coastal management, which would have taken shape in the 1980s; the National Marine Sanctuary Program was sectoral. Considering the list of events upon which Vallejo's analysis is based it is difficult to say when ocean management emerged in policy and law: perhaps UNEP's Blue Plan (1984) for the Mediterranean Sea and the State of Hawaii Ocean Management Plan (1985) can be recorded as landmarks (Figure 9.1).

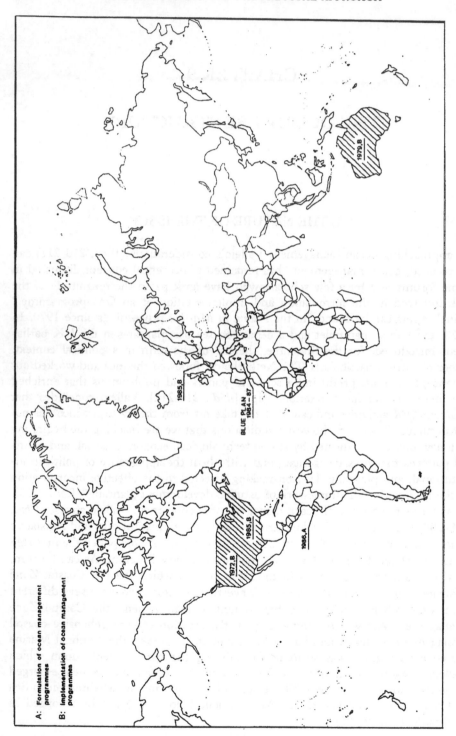

Figure 9.1 *Development and diffusion of ocean management.* Data from Vallejo, 1988, 208-9.

In any case, the consolidation of the concept of ocean management and the legal approach to ocean management may be profitably investigated by the historians of sea management. Here it is convenient to focus one of the most exacting questions, viz what the ocean area is. The answer in the literature is subsequent to the answer given to another question: what the coastal area is. The concept of the coastal area leads to that of ocean area and the delimitation of the former implies that of the latter. In this context what was introduced in Chapter 1 about delimitation criteria is worth examining more deeply.

9.2 THE EXTENT OF THE OCEAN AREA: THE NATURAL ENVIRONMENT

In the present state of the art, in order to define the ocean area, four groups of criteria may be taken into consideration: geomorphological, oceanographic, legal, organisational. Probably geomorphological features were the first to be considered since in the past ocean areas were usually delimited according to the depths of water: the ocean was regarded as the whole marine body extending beyond a given isobath, for instance the 200m one.

Progress in the geomorphological investigation of the continental margin led to the abandoning of these criteria and regarding the outer limit of the (physical) continental shelf as the natural boundary between the coastal sea and the real ocean. According to this criterion the extent of the ocean area varies according to the natural assessment of the continents. For instance, on the eastern side of the central and southern parts of both the Atlantic and Pacific Oceans the (physical) continental shelf is less extensive so the landward limit of the ocean area is near the coastline; on the other hand, in the high latitudes of Europe, North America and Asia the shelf is large, so the landward limit of the ocean area is far from the coastline. More recently attention has been centred also on the slopes and rises and the opportunity of considering these—together with shelves—as components of the coastal assessment has been emphasised. From this stand-point the limit of the ocean management area shifts seawards in some parts of the world where the extent of the slopes and rises is wide: e.g., considering the continental margin instead of the continental shelf, a remarkable seaward displacement of the limit would occur on the south-eastern side of the South America and along the eastern coasts of the United States, and along the eastern coasts of central and southern Africa.

In conclusion, where geomorphological variables—in their turn drawn from the geologic assessment of the seabed—are concerned, three interfaces between coastal and ocean areas may be noted. These are the seaward limits of (i) the continental shelf, (ii) the slope and (iii) the rise. The outer edge of the rise is the most appropriate: there is no doubt that seaward of it the real ocean extends. As a consequence, the development of the rationale in ocean management would require that this edge is the interface between the coastal and the ocean areas.

Oceanographic criteria are provided by the analysis of the physical and chemical properties of the waters and their dynamics. In this context thresholds which the vertical distribution of temperature and salinity of waters pass over are the reference bases for establishing the transition line between the coastal and the ocean areas. In river mouths the outer limit of the fresh-salt water boundary is also considered; as for the dynamics of waters the average low tide-mark is widely considered. In short, some variables already mentioned in Chapter 7 may be taken into consideration again, but this time of course *from the point of view of the ocean area*.

In this context Drozdov's approach [1990], which was discussed in that chapter, is worthy of attention. Taking into consideration both elements and processes—from the aerosol fall to the seabed sediments—ecosystems are identified, partly concerned with the coastal area and partly with the real ocean. Here there is no room for emphasising methodological implications. The results are summarised in Table 9.1.

TABLE 9.1
The geosystems of the coastal area

thermal belts	*cold (polar)*	*warm*	*hot*
environments	*location of the geosystem*		
atmosphere circulation	eastbound transfers in high latitudes	westbound transfers in mid latitudes	monsoon transfers in low latitudes
ocean-continent system		marine	oceanic
river inputs		near river inputs	between river inputs
specific biota		corals mangroves	marches
bottom and coastal land (vertical morphology)		shallow plains	deep-water mountains
coastline (horizontal morphology		fiords rias	lagoons others

According to Drozdov, this taxonomic pattern should be regarded as a preliminary approach. But, particularly bearing ocean management in mind,
methodological implications should be pointed out: (i) geosystems are identified
in as comprehensive a way as possible; (ii) the ocean area is to be delimited by
exclusion, i.e. as a corollary to the establishment of the coastal area. In short,
every environment that is not included in this table should be regarded as pertaining to the ocean. A similar approach could be set up starting *from the opposite
point of view*, i.e. identifying ocean geosystems and grouping these: in this view
coastal geosystems, and subsequent coastal areas, are identified by exclusion.
Anyway it is self-evident that, at the present stage of marine sciences, methodological tools for determining the ocean area as a consequence of the delimitation
of the coastal one are more developed than vice versa.

9.3 THE EXTENT OF THE OCEAN AREA: LEGAL AND ORGANISATIONAL CONTEXTS

As already discussed in Chapter 7, the basic legal criterion distinguishing the
ocean area from the coastal one is made of the distinction between the zones subject to national jurisdiction and areas pertaining to the international regime
(Figure 9.3). From this stand-point both unilateral and bilateral action to determine maritime boundaries [Prescott, 1985, 83–88] and, in particular, the legal
assessment preceding and subsequent to the inclusion of the Exclusive Economic Zone is the basis. Hereunder the alternative views for delimiting the
ocean areas are concisely outlined.

TABLE 9.2
The extent of ocean areas

before the inclusion of the Exclusive Economic Zone in legal frameworks

ocean areas were regarded as extending seawards of the outer edge of the:

territorial sea	contiguous zone	continental shelf
	(much less important criterion, in fact occurring neither in literature nor in planning)	(if claimed or agreed)

after the inclusion of the Exclusive Economic Zone in legal frameworks

ocean areas have been regarded as extending seawards of the outer edge of the:

Exclusive Economic Zone | economic fishery zone (where this zone was claimed or agreed instead of the Exclusive Economic Zone)

In conclusion, from the legal stand-point, ocean areas extend seaward from (i) the outer limit of the territorial sea, where neither the continental shelf nor the Exclusive Economic Zone have been claimed, (ii) the outer limit of the continental shelf, where only this belt has been claimed or agreed, (iii) the outer limit of the Exclusive Economic Zone, where only this belt has been claimed: (iv) the most seaward advanced jurisdictional limit, where both the continental shelf and the Exclusive Economic Zone were claimed or agreed.

As far as organisational patterns are concerned, attention is centred on the way in which human communities conduct activities on the sea, such as setting up use structures, exploring and exploiting resources, and developing navigation. When these elements provide the criteria to establish the extent of the coastal area and, as a result, to identify its boundary with the ocean area, planners take into account the kinds of structures and other installations by which land facilities are extended seawards. Marine areas in which facilities physically linked to the land are located, such as port structures, offshore loading or unloading terminals, offshore power plants, desalination plants, offshore oil and gas installations, artificial islands, floating settlements, subsea settlements, etc, give form to the coastal area: beyond it extends the ocean space.

The aforementioned framework of criteria is too fragmented to allow us to set up comprehensive and as objective as possible criteria for the delimitation of ocean areas. Three deductions seem appropriate to highlight the current state of the art.

(i)   As far as *natural criteria* are concerned, whichever of these is applied, it is possible to state that the interface between the ocean area and the coastal one is not located landward of the outer edge of the continental margin.

(ii)   Wherever the continental margin is narrow—which occurs in many parts of the world—*legal criteria* lead to establishing the ocean limit much further to seaward than natural criteria do. This depends on the need to ensure marine areas as large as possible for national jurisdictions: it is a consequence of the propensity sustained by the 1982 Convention.

(iii) *Organisational criteria* lead one to think that the ocean area extends where industrial plants are not physically linked—through pipelines, conveyors belts, etc.—to the land. In this view the coastal area appears narrow because, only in a few parts of the world, offshore oil and gas platforms—i.e. the installations that, by their nature, are mostly extending seawards—are located in places quite distant from the coastline. As a consequence, from the organisational point of view the ocean areas appear much more extended than they are according to legal criteria.

9.4 THE ISLAND WORLD

Islands and archipelagoes are a separate concern in this complicated framework of delimitation criteria and subsequent implications for management patterns. These are a special issue because the relationships between the natural environ-

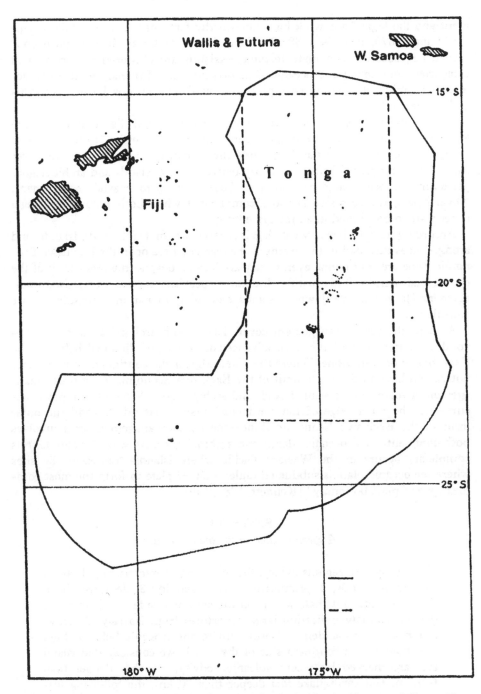

Figure 9.2 *The national jurisdictional extent of oceanic islands: the case of Tonga.* From Buchholz, 1987, 86.

ment and the legal framework have noteworthy features [Prescott, 1985, Chapter 7, and 1987b; Briscoe, 1988; Buccholz, 1987]. As far as their origin is concerned-i.e., according to plate tectonics—islands are clustered into two main categories: continental and oceanic. The present state of management is characterized by many interesting examples, such as the Galapagos Islands [Broadus and Gaines, 1987].

Continental islands belong to the continental margin while oceanic ones are located in ocean areas, i.e. beyond the outer edge of the margin. If the ocean area is identified according to natural criteria the response is very simple: the outer edge of the margin marks the boundary between those islands and archipelagoes pertaining to ocean management and those relating to coastal management. Nevertheless, the response is more complicated when both natural and legal criteria are to be applied in an integrated way.

According to the latest view, legal criteria are first followed: islands and archipelagoes included in the extent of the territorial sea or in the Exclusive Economic Zone are continental even if they are located beyond the outer limit of the continental margin. This justifies literature speaking *tout court* of coastal management [Johnson, 1989] when islands belonging to national jurisdiction are considered.

As a result, ocean management covers islands and archipelagoes located beyond the outer limit of the most advanced national jurisdictional belt, i.e. the Exclusive Economic Zone. Where the outer edge of the continental margin does not extend beyond the outer limit of the Exclusive Economic Zone ocean management involves only oceanic islands and archipelagoes. Where the opposite occurs—i.e. the outer edge of the continental margin extends beyond the outer limit of the Exclusive Economic Zone—the ocean management area involves both continental and oceanic islands and archipelagoes. By way of example, this complexity occurs in the Western Pacific where islands "represent the sites where the oceanic plate is subducted under another plate to form the most complex type of plate boundary" [Kennett, 1982, 149].

Subject 9.1
A special island area: the island arc

"The island-arc regions exhibit the following characteristics: 1. arcuate line of islands; 2. prominent volcanic activity; 3. deep trench on the ocean side and shallow seas on the continent side; 4. distinct line gravity anomaly indicating large departures from isostasy; 5. active tectonism; 6. coincidence of arcs with recent orogenic belts; 7. high heat flow on the continent side of the arc. If we consider that island arcs are marked by recent volcanic activity, trenches deeper than 6000 m, and earthquake foci deeper than 70 km, the following features are island arcs: (1) New Zealand to Tonga; (2) Melanesia; (3) Indonesia; (4) Philippines; (5) Formosa (Taiwan) and western Japan;

(6) Marianna and eastern Japan; (7) Kurile and Kamchatka; (8) Aleutian and Alaska; (9) Central America; (10) West Indies (Lesser Antilles); (11) South America; and (12) Scotia Arc and West Antarctica". [Kennett, 1982, 151].

The role of Pacific oceanic islands in the management context is worth pointing out. Most of them should be regarded as pertaining to ocean area management because the geological and geomorphological contexts, as well ecosystems, are part of the real ocean world. Nevertheless, they are covered by Exclusive Economic Zones so that, from the jurisdictional point of view, they develop *coastal* area management in a large surrounding *ocean* space. This feature is important in archipelagoes: according to the criteria for tracing baselines provided by the 1982 Convention (Articles 46 and 47) large marine spaces belong to (legal) archipelagic waters: as a result, the Exclusive Economic Zone, since it extends from the outer limit of these area, covers very large ocean spaces. The determination of baselines and subsequent claims to Exclusive Economic Zones not only has radically changed the economic and political geography of the Pacific but has also drastically restricted the spaces subject to the international régime and, as a consequence, to ocean management patterns. Prescott's analysis [1985, Chapters 7 and 9] demonstrates how profound these implications are.

Subject 9.2
The physical structure of oceanic islands: the Hawaiian case

"The classic example of a systematic age progression along a linear chain is the Hawaiian-Emperor-Chain. The Hawaiian Islands lie on the Hawaiian Ridge, which runs for about 2600 km from the island of Hawaii to the coral atolls of Midway and Kure. Northwest of Kauai, the ridge continues for another 2000 km, mainly as a submarine ridge. Nihoa and Necker represent eroded stumps of volcanoes just above sea level. After the Hawaiian-Emperor bends, it strikes northnorthwest toward the Meiji Guyot (...) Situated on the island of Hawaii at the southeastern limit of the chain are several active volcanoes. Kilauea is the most, while Manua Loa and Hualalei are less active (...)". [Kennett, 1982, 163].

9.5 OCEAN MANAGEMENT: MAIN GOALS

The thesis supporting the present analysis is the rationale of developing ocean management in marine spaces extending seawards from the outer edge of the continental margin. Further, society, which is pragmatic and resource-oriented, is already considering the outermost limit of national jurisdictional zones as the boundary between coastal and ocean areas. Starting from this basis reasoning can first deal with the opportunities, constraints and implications arising from the international régime and then with the goals of ocean management.

As far as the first step, namely, the legal framework, is concerned, the following aspects seem worthy of consideration.

(i)   *High seas.* According to Part VII of the 1982 Convention, high seas—i.e. "all parts of the sea that are not included in the Exclusive Economic Zone, in the territorial sea or in the internal waters of a State, or in the archipelagic waters of an archipelagic State" (art. 86)—belong to the "freedom ocean". These spaces "are open to all States, whether coastal or land-locked" (art 87.1), which, *inter alia*, have equal rights in navigation, overflying, fishing, conducting scientific research, laying submarine cables, constructing artificial island and other similar installations. As a result, high seas are regarded as the portion of the sea in which, at the present time, Grotius's principle of the freedom of the seas is put into practice—or, better still, is guaranteed.

(ii) *Deep seabeds.* According to Part XI of the 1982 Convention, the seabed and subsoil extending beyond the marine areas under national jurisdiction—i.e. deep seabeds [Prescott, 1985, 124-138]—make up the *Area.* "The Area and its resources are the common heritage of mankind" (art 136) and the International Seabed Authority should promote and manage co-operation, especially along the north-south co-ordinate, for exploiting mineral resources and protecting and preserving the ocean environment. It is worth evaluating that, while the legal status of the water column and surface (high seas) aims at guaranteeing ocean uses to every state, the objective of the deep seabed consists of using it through a supra-national body. As noted in Chapter 4, the extent of high seas and deep seabed do not coincide.

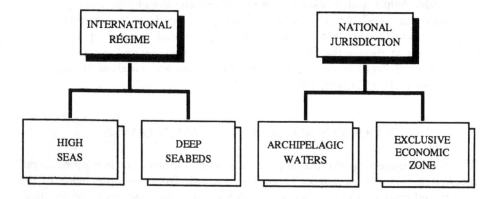

Figure 9.3 *International régime and national jurisdictional zones involving the oceanic space* (extending beyond the continental margin).

(iii) *Archipelagic waters*. Although the archipelagic waters are under national jurisdiction in these spaces, aliens may exercise rights: in addition to the innocent passage and overflying rights, they can maintain fishing activities exercised prior the establishment of the archipelagic marine area, as well as submarine cables, and can develop research with the consent of the archipelagic state. At first, it might appear that, because of the large extent of archipelagic waters, the international community has agreed to preserve some features of the international régime but, as a matter of fact, freedom in archipelagic waters is very small: it is similar to that of the territorial sea.

(iv) *General conventions*. As far as environmental protection of the ocean as a whole is concerned, general conventions, such as MARPOL (Convention for Prevention of Marine Pollution by Dumping from Ships and Aircraft, 1972), have come into force. These are applied not only to the high seas but also to marine spaces subject to national jurisdiction, but in the former case they have particular importance since they are the main background for ocean management.

(v) *Regional conventions*. In order to manage the ocean environment *regional conventions* have acquired increasing importance, for at least two reasons. First, they are usually provided with protocols regulating the pollution sources in a very detailed way and in efficient co-ordination with general conventions. The Barcelona Convention (1976) for the Mediterranean Sea is a good example. Secondly, they are related to a marine space, e.g. a semi-enclosed sea, in its entirety including both area subject to national jurisdiction and high seas. As a consequence, they ensure homogeneous environmental management, or at least they open the way to this goal.

On the basis of this framework the following set of management goals may be considered:

1.   *development-oriented objectives*
1.1 the implementation of conventional ocean uses, such as navigation and fishing, under the umbrella of high seas' freedoms;
1.2 the setting up of new ocean uses, such as renewable energy production and recreational uses, also under the umbrella of high seas' principles;
1.3 the setting up of new ocean uses by the exploitation of the deep seabed and subsoil under the umbrella of the Area policy and subsequent international co-operation;
1.4 the minimisation of conflicts between states either in surface and water column uses and in the expected deep seabed and subsoil uses;

2.   *environmental management*
2.1 the protection and preservation of the marine ecosystem;
2.2 actions for special areas and particularly sensitive areas;
2.3 the conservation of special ecosystems and species;
2.4 the protection of the seabed.

TABLE 9.3
The ocean use framework

categories	sub-categories
2. SHIPPING, CARRIERS	2.1 bulk vessels
	2.2 general cargo vessels
	2.3 unitized cargo vessels
	2.4 heavy and large cargo vessels
	2.5 passenger vessels
	2.6 multipurpose vessels
3. SHIPPING, ROUTES	3.1 mid-sea and deep-sea routes
4. SHIPPING, NAVIGATION AIDS	4.3 hyperbolic systems
	4.4 satellite systems
	4.5 inertial systems
5. SEA PIPELINES	5.2 liquid bulk pipelines
	5.3 gas pipelines
6. CABLES	6.2 telephone cables
8. BIOLOGICAL RESOURCES	8.1 fishing
9. HYDROCARBONS	9.1 exploration
10. METALLIFEROUS RESOURCES	10.3 seabed deposits
11. RENEWABLE ENERGY SOURCES	11.2 water properties
12. DEFENCE	12.1 exercise areas
	12.2 nuclear test areas
	12.4 explosive weapon areas
13. RECREATION	13.2 offshore facilities and activities
15. WASTE DISPOSAL	15.4 dumping
	15.5 navigation
16. RESEARCH	16.1 water column
	16.2 seabed and subsoil
	16.3 ecosystems
	16.4 external environmentinteraction
	16.5 special areas and particularly sensitive areas
	16.6 ocean management
18. ENVIRONMENT PROTECTION AND PRESERVATION	18.2 offshore

9.6 THE OCEAN USE STRUCTURE: USE FRAMEWORK

At this point of reasoning, which leads to the analysis of the ocean use structure, a *mise au point* should be made. As has been stated, the ocean area can be conceived as delimited in three ways:

i)   as the area extending seaward from the average low water mark or geomorphological-oceanographic thresholds;

(ii) as the area extending seaward to the outer edge of the continental margin;

(iii) as the area extending seaward to the outer limit of the widest national jurisdictional belt (Exclusive Economic Zone).

However the extent of the ocean area is conceived, analysis relating to ocean management has three stages dealing with: (a) the framework of ocean uses; (b) the relationships between ocean uses, (c) the relationships between ocean uses and the ocean environment. This approach is similar to that applied for the coastal use structure (Chapter 7). According to the general approach developed in Chapter 5, the *ocean use framework* may be formulated as shown in *Appendix E*. Only the categories and sub-categories of sea uses more or less tied to the ocean environment are grouped below.

9.7 THE OCEAN USE-USE RELATIONSHIP MODEL

Beginning with this framework the relationships among ocean uses can now be considered. An approach similar to that applied to analysis of the coastal area can be followed, grouping ocean use-use relationships in the following terms:

1.   *non-existence of use-use relationships*

2.   *existence of relationships*
2.1 neutral relationships
2.2 conflicting relationships
2.3 beneficial relationships.

As far as the *vector of the relationship* is concerned, three kinds of inputs are evident: (i) those moving from use i to use j, (ii) those moving from use j to use i and (iii) those moving simultaneously from i to j and *vice versa*. Where

O_i, are the ocean uses included at matrix right, i being ranged from 1 to n;

O_j, are the coastal uses included at matrix left, j being ranged from 1 to n ;

the following cases can occur

1) $O_i = f(O_j)$

2) $O_j = f(O_i)$

3) $O_i = f(O_j); O_j = f(O_i).$

In cases (1) and (2) simple patterns of relationships occur, which are substantially referrable to the model of series relationships. Case (3) is too complicated to be referred to a given pattern of relationships.

Many coastal areas are involved in as many uses and with as many users as to justify evaluating the differentiation of uses and the co-ordination between the ways in which they are managed. As a consequence, the differentiation-coordination model appears as a useful tool. On the other hand, ocean areas are commonly not involved in many uses, so the differentiation of uses is very low. Also the co-ordination of uses is much easier to achieve than in the coastal area. High differentiation and the risk of having little co-ordination can occur in ocean areas involved in several kinds of navigation (mercantile, naval, recreational) and in well developed fishing. These considerations allow us to believe that the differentiation-coordination model is less significant in ocean areas than in the coastal ones. However, it is not out of place to include this model as a conceptual tool for the analysis and management of these areas. The main reason for this is due to the growth of sea uses beyond the legal boundaries of coastal areas in spaces subject to the international régime.

As far as management is concerned, the category of conflicting relationships is worthy of special attention in order to find methods to minimize tensions between ocean uses. In this context four categories of conflicting relationships can be considered:

(i)   two or more uses are located in the same place, bringing about *locational incompatibility*; e.g., conflicts between research and naval exercise areas;

(ii)   one use is harmful to another giving form to *organisational incompatibility*; e.g., conflicts between naval activities—particularly in areas characterized by the presence of superpower fleets [Barnaby, 1985]—and yacht racing and cruising, and *sensu lato* recreational uses [Anderson, 1980];

(iii) one use causes environmental impacts that the other cannot tolerate, producing *environmental incompatibility*; e.g., the changes in water temperature brought about by Ocean Thermal Electric Plants can damage the marine environment, the chances of hazardous material incidents [Bertram and Santini, 1988] and the impacts due to radioactive material dumping [Bewers and Garrett, 1987; Frosh, Hollister and Deese, 1978; Van Dyke, 1988] jeopardize the prospects of exploiting biological resources;

(iv)   one use give form to a visual impact that the other cannot bear; e.g., this *aesthetic incompatibility* could occur when deep sea mining creates a *seascape* that will be regarded as not aesthetically appropriate for recreational navigation.

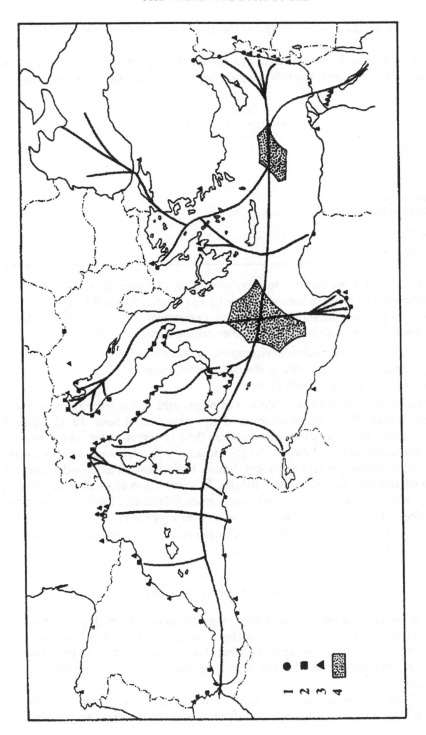

Figure 9.4 *Oil and gas industry in the Mediterranean Sea*: 1) loading port for crude oil; 2) unloading port for crude oil; 3) refinery; 4) area where the discharge of oil is permitted.

9.8 THE OCEAN USE-ENVIRONMENT RELATIONSHIP MODEL

Shifting attention from the relationships between ocean uses to those between uses and the ecosystem, the following components of the ocean area emerge:

TABLE 9.4
The components of the ocean environment

---

sea surface

water column:
upper layers
intermediate layers
lower layers

seabed

subsoil

From this starting point a matrix, similar to that pertaining to coastal area management can be sketched, in which (i) the lines embrace the components of the ocean environment and (ii) the columns encompass ocean uses. The relationships between the two ranges of variables can be clustered into three categories:

(i)   inputs from the ocean environment to ocean uses;
(ii) inputs from ocean uses to the ocean environment;
(iii) inputs moving in both directions.

At the present stage of development of ocean uses, a smaller number of uses in the ocean area affect the environment than in the coastal zone. In addition, the level of pollution in ocean areas varies less than in coastal zones where large annual variations occur. Environmental impacts in ocean areas arise especially from navigation—particularly from liquid bulk transportation—defence, offshore oil and gas exploitation and fishing. In some ocean areas they are so sparse as to justify speaking of the sea as a waste space [Goldberg, 1985]. In this context attention is focused upon new developments in ocean uses, which may be rapid because of the present remarkable technological advance and the diffusion may be large because of economic strategies.

Subject 9.3
The prospects for Pacific mining

As is well known, the most important advantages from deepsea mining are expected to derive from the exploitation of Pacific resources.
"All major seafloor fracture systems should be considered as potential sites for marine sulphide deposits, as should the areas adjacent to such systems.
Technological advances in exploration will greatly accelerate the rate of exploration of prospective seafloor areas over the next two decades.

The implication is that some of the best deposits will be discovered during that period. Early support of exploration enhances the long-term chances of participating in commercial marine sulphide or crust-mining operations.

Which state or states will lead in the future commercial development of seabed mining is not yet clear. A few possible combinations are (1) Japan or a consortium including Japan and one or more other Asian countries, (2) West Germany or a consortium including West Germany and other West European countries, or (3) a consortium including Japan, the United States, and West Germany.

Active involvement of China and India—and perhaps other large developing countries—in exploration of seabed minerals is likely". [Johnson and Clark, 1988, 158].

TABLE 9.5
Relationships between naval uses and the ocean environment

12 DEFENCE / ENVIRONMENT	12.1 exercise areas				12.2 nuclear test areas	12.4 explosive weapon areas			
	12.1.1 air firing exercise areas	12.1.2 surface firing exercise areas	12.1.3 surface bombing exercise areas	12.1.4 submarine exercise areas		12.4.1 mines	12.4.2 torpedoes	12.4.3 depth charges	12.4.4 missiles
sea surface	■	■	■		■		■		■
water column — upper layers	■	■	■		■		■		■
water column — intermediate layers			■	■	■			■	■
water column — lower layers				■	■	■		■	■
seabed				■	■	■		■	■
subsoil				■	■	■		■	■

▲ inputs from the environment to ocean uses

■ inputs from ocean uses to the environment

● inputs in both directions

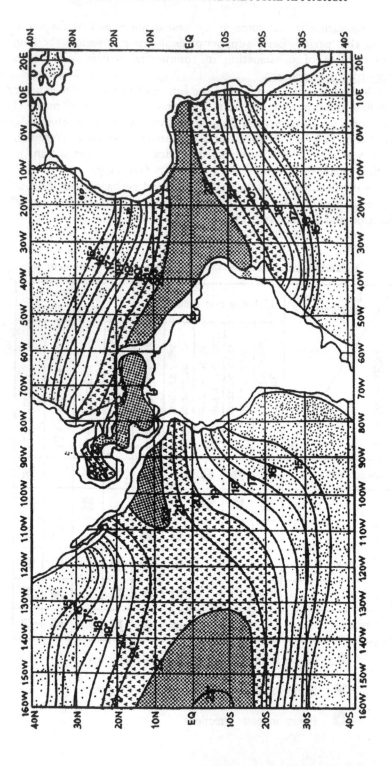

Figure 9.5(top) *Distribution of OTEC resources* between the sea surface and 1,000 m depth. From UN DIESA, 1984, 3.

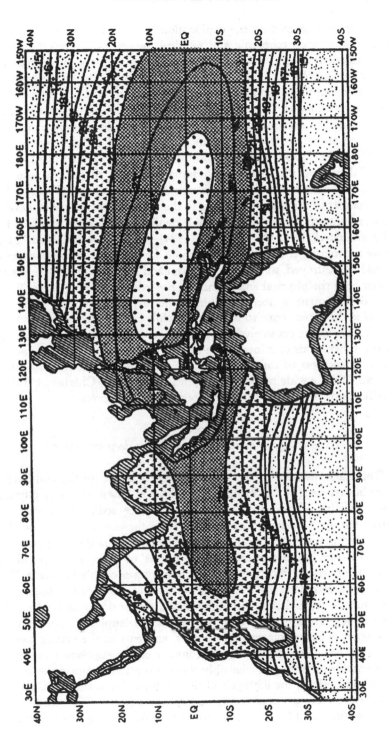

Figure 9.5.(bottom) *Distribution of OTEC resources* between the sea surface and 1,000 m depth. From UN DIESA, 1984, 3.

First, interest is centred on the exploitation of renewable ocean resources. It is well known how wide economic and political interest is in exploiting the ocean thermal gradient through the OTEC (Ocean Thermal Energy Conversion) plants [Ford, Niblett and Walker, 1987, 61-78]. The economic objective consists of supplementing conventional energy production (from oil, coal and nuclear sources) with the new technique. Secondly, as research developed in the context of the United Nations has demonstrated, this renewable source is easily exploitable in a semi-enclosed space, so it could be an important tool for developing countries [UN DIESA, 1984]. Both scientists and technologists agree that this technology is capable of damaging the marine environment because there is a suspicion that thermal operations can alter the ecosystem of the water column (Figure 9.5).

In the context of minerals, according to the technologies that have developed to date, the exploitation of manganese nodule deposits—the extent of which has been carefully evaluated [UN, OETB, 1982, 1984]—is suspected of being capable of damaging the marine environment as a whole, i.e. from the seabed, where ecosystems could be destroyed, to the water column and the sea surface where sediments could alter both the physical and chemical properties of waters (Figure 9.6). However, the development of deepsea mining is not expected to occur as early as was thought of in the recent past, so in the meanwhile new technologies capable of preventing damage to the ocean environment can be developed. Of course, in addition to nodules, other kinds of minerals can be provided by ocean seabeds and subsoils; from the oil and gas to micronodules and metalliferous sediments [Blissenbach and Nawab, 1982] a wide range exists [Kent, 1980, 62–67; Charlier, 1983] generating specific needs for environmental protection and preservation.

Subject 9.4
Expected environmental impacts from deep sea mining

"The major menace to the water column is posed by the continuous line bucket system that brings the sea floor sediment in direct physical contact with all ecological ocean layers, since sediment washes out from the open buckets during ascent from the ocean floor to the ship (...). Air or water-lift mining systems results in a surface discharge of nutrient-rich bottom water causing a 'plume' of sediment to spread out over the ocean surface inducing the following probable effects: a. The plume spreads laterally over large surfaces of the ocean, discolouring the surface water before it is diluted and settles downward (...). b. In the upper layers of the ocean, known as the 'euphotic zone', phytoplankton exists, photosynthesis occurs, and part of the earth's oxygen is produced. The depositing of sediment in the euphotic zone limits penetration of sunlight, plausibly threatening the photosynthetic process and perhaps the life cycle of the phytoplankton, thus endangering one of the early stages of the ocean food chain (...). c. Filter feeding fish such as tuna could also be expected to be influenced by the presence of

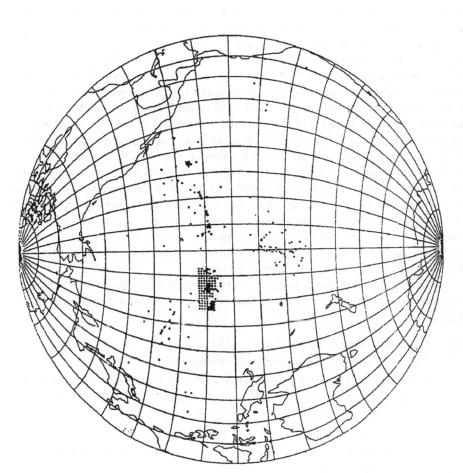

Figure 9.6 *Pacific areas where nodule manganese deposits have been measured* (kg/sq m). From UN OETB, 1982, 38-9.

a plume, although it is as yet unknown whether the effects would be deleterious or beneficial (...). d. Floor sediment contains dormant spores that might be reactivated to the reproductive cycle in the warmer and lighter waters of the ocean surface". [Post, 1983, 59].

Bearing the present and the expected spectrum of ocean uses in mind it can be stated that the *exploitation-change model* discussed in Chapter 5 does not have a significant role in explaining and investigating the environmental impact in marine spaces extending beyond the outer edge of the continental margin or beyond the outer limit of the Exclusive Economic Zone—in short, beyond the outer limit of what is thought of as the coastal area. In fact, this model, when is related to the coastal area (Chapter 7), helps us to reason about the relationships between the growth of coastal uses—particularly that of harmful and hazardous uses—and the increase of the pollution level in the marine environment. On the contrary, where the ocean area is concerned, the growth of the uses is too slow to suggest that it is able to lead to collapse of the ocean environment. This morphogenetic event, i.e. collapse, can occur because of catastrophic events due to the environment itself—e.g. volcanic eruptions—or to social behaviour, e.g. a collision of tankers or a disaster in powered carriers. As a consequence, it must be said that, in the present historic stage, the ocean area undergoes adjustments and can collapse only because of unexpected events. These events are to be prevented of course, but—as far as management is concerned—they cannot be regarded as normal. The evolution of the ocean area consists of the transition of the ocean structure through adjustment phases.

In this theoretical context the equation of the state of the ocean area can be introduced. The approach is similar to that relating to coastal area management.

Where

1) O_{to} is the state of the ocean area in the time t_o;

2) O_{t1} is the state of the ocean area in the time t_1:

3) u is the set of parameters, depending on the nature of relationships between ocean uses, which bring about the changes of the ocean structure;

4) u_{to} is the set of these parameters in the time t_o;

5) u_{t1} is the set of these parameters in the time t_1,

according to the general theory of the equation of the state, it is stated that

6) $O_{t1} = F [O_{to}, u_{to}; u_{t1}]$

This approach merits the same comments that were formulated in Chapter 7, when the equation of the state was referred to the coastal area. In particular, *in its nature* this equation is deterministic because it implies that O_{t1} is determined by O_{to}, i.e. the future is the product of the present. In order to prevent this inappropriate approach the equation of the state should be integrated by non-quantitative analyses of the evolution of technologies, organisational patterns and the social perception of coastal management.

CHAPTER 10

MANAGING THE OCEAN AREA

10.1 THE OCEAN MANAGEMENT SYSTEM

To some extent the starting point from which to approach the ocean management system could be similar to that of coastal management (Chapter 8), since ocean areas also have to be considered through three kinds of variables: law, decision making centres, and decision making patterns. When the ocean area is delimited by *legal criteria* and is regarded as the marine space extending beyond the outer limit of the Exclusive Economic Zone, the least complicated management system applies consisting of:

Legal framework
(i)   general conventions;
(ii) regional conventions;

Goals
(iii) resource management;
(iv) environmental management;
(v)   research;

Decision making centres
(vi) inter-governmental decision-making centres, considered in relation to their sectoral competence, and their "ratione loci" as well;
(vii) extra-governmental decision-making centres, considered according to their operational fields;

Decision making patterns
(viii) decisional patterns applied by inter-governmental centres;
(ix)   decisional patterns applied by extra-governmental centres.

A separate case occurs when both the continental shelf and the Exclusive Economic Zone are claimed and the former is wider than the latter. In this case a zone ranged between the outer limit of the Exclusive Economic Zone and the outer edge of the continental margin is included in the ocean area but only its seabed and subsoil are covered by a national jurisdictional zone. As a result, the whole ocean environment extending beyond the outer edge of the continental margin plus the water column and sea surface of the zone extending from this boundary and the outer edge of the Exclusive Economic Zone, are of international status.

When the ocean area is conceived through ecosystem- and physical environment-based criteria, its identification is more justified than it is when it is thought of in terms of legal criteria. This is due to the fact that natural criteria lead to distinguishing the real ocean environment, which in its nature calls for specific management criteria, from the coastal environment, which in its turn calls for other kinds of criteria. Nevertheless, management depends above all upon legal factors because the possibility of setting up ocean management exists only in those areas covered by the international legal régime. The most important implication of this occurs where the continental margin is less extensive than the Exclusive Economic Zone. In this case the rationale would be that, beyond the outer edge of the continental margin, ocean management systems are practised. But it cannot occur because the zone extending between the outer edge of the continental margin and the outer limit of the Exclusive Economic Zone is subject to national jurisdiction. As a consequence, ocean management is based both on international régime (areas extending seaward the outer limit of the Exclusive Economic Zone) and national law. This occur in many parts of the ocean world, as can be seen by comparing the extent of the continental margin with the 200 nm boundary (Chapter 4, Paragraph 5). In these zones a potentially more complicated framework than that mentioned above occurs.

Legal factors
(i)   general conventions;
(ii)   regional conventions;
(iii) national law;

Goals
(iv)   resource management;
(v)   environmental management;
(vi)   research;

Decision-making centres
(vii) governmental decision making centres;
(viii) inter-governmental decision making centres;
(ix)   extra-governmental decision making centres.

Decision-making patterns
(x)   patterns from governmental centres;
(xi)   patterns from governmental centres;
(xii) patterns from extra-governmental centres.

As can be seen, the ocean management concept is ambiguous because (i) the ocean area can thought of, and delimited as well, by means of two alternative criteria, legal and natural, and (ii) there does not exist any appropriate way to apply both these criteria in such a way as to make a comprehensive approach. The *theoretical approach* looks at the environment, so it is based on the consideration of ecosystems and their physical environments and leads to assuming environmental protection and preservation as the main goal of management: it is holistic. The *legal approach* is based on the division of the sea into two main areas, provided respectively with national and international status. Bearing in mind that, at the present stage, sea management is primarily conducted with reference to criteria drawn from the law of the sea, three circumstances seem worth noting:

(i)   the resources of the sea surface and water column pertains to the status of high seas and are to be managed respecting the subsequent principle of freedom;

(ii) the resources of the seabed and subsoil are regarded as the "common heritage of mankind" and are to be referred to the régime of the Area;

(iii) for environmental management, the 1982 UNCLOS Convention tends to create a code of conduct to act as an umbrella for general and regional conventions.

When attention shifts from law and decision-making systems to the goals of ocean management systems, three categories can be considered: resource management, environmental management, and research. This framework emerges.

Resource management
Relevant components of the ocean environment:
(i)   sea surface and water column
(ii) seabed and subsoil

Environmental management
Main factors of pollution or disturbance:
(iii) land-based sources
(iv)   exploitation of coastal seabed and subsoil resources
(v)   exploitation of ocean seabed and subsoil resources
(vi)   dumping
(vii) vessels
(viii) atmospheric sources
(ix)   archaeological and historical finds

Research
Objective of exploration and other research activities:
(x)   implementing ocean knowledge
(xi)   exploiting resources
(xii) protecting the environment.

Subject 10.1
Management principles from the 1982 Convention

As for the goals of ocean management considered in the above framework the following articles of the 1982 Convention can be regarded as
reference points.
– archaeological and historical objects, art. 149
– pollution from land-based sources, art. 207 and 213
– pollution from activities on coastal seabeds, art. 208 and 214
– pollution from ocean seabed and subsoil activities, art. 145, 209
  and 215
– dumping, art. 210 and 216 navigation, art. 211 and 217-221
– pollution through atmosphere, art. 212 and 222

There is no room to formulate a detailed framework for the decision centres
involved in ocean management. However, some remarks seem to be appropriate.
The potentially most important and powerful international body concerned
with ocean affairs is the International Seabed Authority whose tasks consist of
the exploitation of mineral resources of the Area and subsequent environmental
protection. The literature on the law of the sea has pointed out the main patterns
through which the Authority is expected to act: (i) the co-ordination between
states exploiting manganese nodule fields on deep seabeds; (ii) the exploitation
of these fields through the Enterprise in which co-operation between developed
and developing countries should be implemented at the operational level [Post,
1983, Chapter 7]. Nevertheless, according to current opinion it is less probable
that this management system will be established whatever vicissitudes the Convention undergoes.
Current international bodies involved in ocean management do not exercise
real authority, in the sense that they do not act as decision-making centres in the
management of resources and environment. Their operational fields include
doing research, promoting co-operation, and helping States and local bodies.
Nevertheless, they have acquired growing importance to the point of believing
that, when ocean management is as important as the coastal management
presently is, they could be the best bases to apply comprehensive managerial criteria. In this context the FAO for biological resources, UNEP for environmental
management and IMO for navigation safety and subsequent environmental implications may be regarded as the main bodies concerned.
In spite of its international status the ocean area also calls for action from
states. This action is especially needed for protecting the environment because it
is up to coastal, island and archipelagic states to prevent land-based sources and
coastal seabed activities from polluting ocean areas. Inter-state co-operation taking place in the context of the regional conventions is the best tool for achieving
this objective.

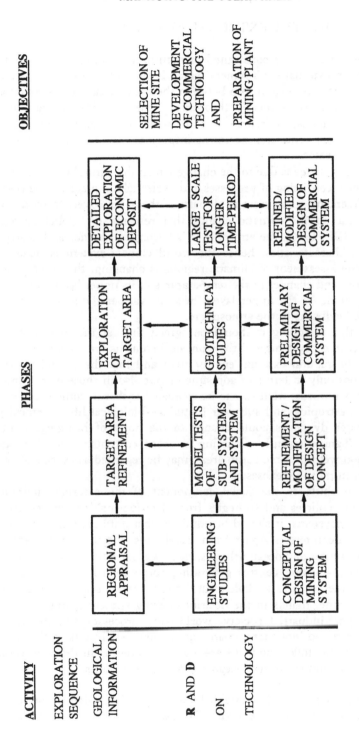

Figure 10.1 *Deep sea mining: technology development and information collection in exploration sequence.* From UN OETB, 1984, 17.

10.2 THE EXTERNAL ENVIRONMENT

The ocean use structure is constituted of natural and human components and interacts with an external environment which is similarly constituted. As a preliminary approach, the relationship between the ocean area and the external environment seems to be a merely theoretical issue with insignificant practical relevance. Nevertheless, the relationships between the ocean area and its external environment are acquiring increasing importance as the human pressure on the high seas increases.

First, this importance is due to the change which the natural external environment undergoes, consisting of processes and cycles taking place on the planetary scale, and influencing ocean areas in all parts of the ocean cover. In particular, its influence is due to the climatic change that research on global change has focused on [IGBP, 1990]: the atmosphere is becoming warmer and is acquiring new chemical characteristics, hence the world climate system is being transformed and the assessment of climatic regions is changing, the ocean is subject to sea level rise, and changes in the temperatures of its upper layers, as well as in its current system. As the upper layers become warmer the air-sea interface is affected, with feedbacks to the atmosphere.

Following this reasoning it should be agreed that, at the present historical stage, both ocean ecosystems and their physical contexts are receiving important inputs from the natural external environment and that they could respond to these inputs not only in terms of adjustments but also in those of real morphogeneses. In other words, what it is to be evaluated is that in some ocean zones—both at the intertropical and subpolar latitudes—biological life, chemical properties and water dynamics could change to the point of bringing about new ecosystems. Changes recently occurring in ocean currents, which have deeply influenced food webs in some ocean areas, may be regarded as aspects of changes leading to ocean morphogeneses.

As attention shifts to the social component of the external environment, namely national policies and strategies from decision-making centres involving large marine ecosystems, it should be admitted that until recently this issue did not deserve wide interest. Apart from naval uses, maritime merchant traffic along deep sea routes and ocean fisheries, there was no action coming from international decision-making centres generating comprehensive management patterns either in the ocean as a whole or at the sub-oceanic scale. On the contrary, at the present time, thanks to the decision-making strategies and technological advances of post-industrial society, world-wide processes acquire more and more importance and influences ocean management in all parts of the ocean. As a first approach, the following fields are worth considering in the prospect of the establishment of comprehensive ocean management patterns.

(i)   *Biological resources.* The establishment of the Exclusive Economic Zones [Lucas and Loftas, 1982] and Exclusive Fishery Zones have brought

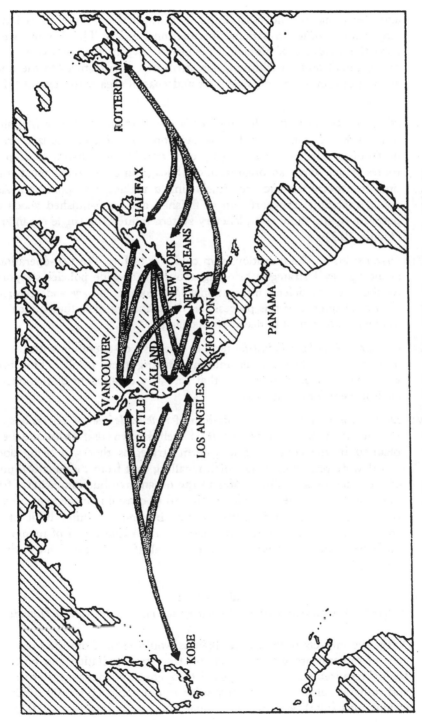

Figure 10.2 *The network of landbridges in the 1980s.* From Hayuth, 1982, 16.

about—as one of the most important implications—the inclusion of large fishery zones in national jurisdictional zones. This has not weakened the weight that biological resources have in the context of ocean management and, in addition, has attributed new features to the interaction between ocean management and coastal area management [Holt, 1978].

(ii) *Maritime transportation* has entered a phase of great change [Abrahmsonn, 1983; Frankel, 1989]. The development of containerized transportation has led to the setting up of a world-wide network of deep sea routes which form an organizational background involving most maritime transportation and are closely linked to land transportation. As a result, maritime transportation of finished and semi-finished goods appears more and more as a planetary system influencing single ocean areas (Figure 10.2).

(iii) *Ocean stresses*. The transition from cold war to East-West co-operation is changing naval strategies bringing about world-wide policies supported by the need to defend international law, combat piracy and terrorism. As a consequence of this process, on the planetary scale the political and military involvement of ocean areas changes.

(iv) *Offshore oil and gas industries* tend to establish co-operation with the aim of minimizing their efforts and maximizing their results. This bring about world-wide co-operation that acts as a component of the external environment for ocean areas.

(v) *Deep sea mining* and the exploitation of *renewable energy sources* has still not involved ocean areas but it is well known that these are being planned in the context of long-term strategies that will bring about world-wide processes. It is worth recalling that from the practical point of view deep sea mining relates to the ocean area but somewhere, from the legal point of view, it can be referred to coastal management because the Exclusive Economic Zone extends beyond the outer limit of the continental margin. By way of example, the exploitation of manganese nodules in the Hawaiian area is a coastal affair of the United States [Holtz, 1988].

Subject 10.2
Expected organisational patterns from deep sea mining: the case of Hawaii

"Negative impacts to the coastal regions would result from the presence of mineral processing plants and from transportation and port facility requirements. The proposed throughput of manganese crust and mineral substrate would require a mining ship of 65,000 dead weight tons (dwt), and two additional bulk cargo vessels of 30,000

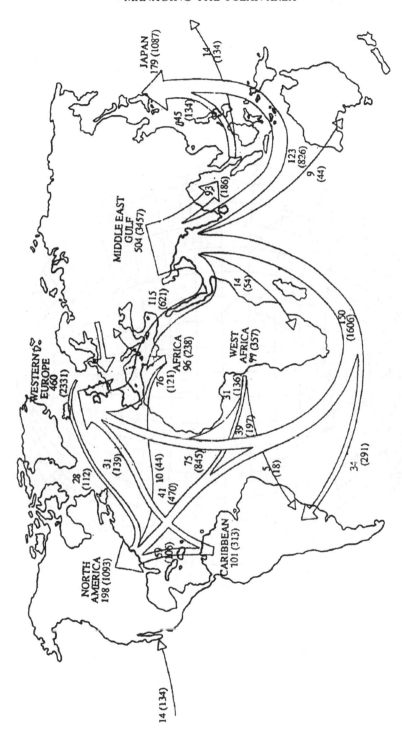

Figure 10.3 *Maritime transportation of crude oil*. From Fearnleys, World Bulk Trade (late 1980s): million tons and (between brackets) billion tons × miles.

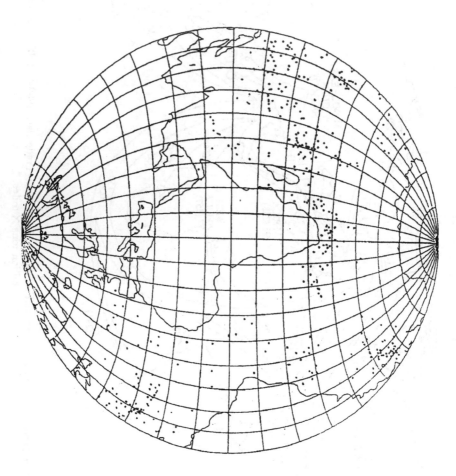

Figure 10.4 *Atlantic and Indian Ocean locations from where manganese nodules and crusts have been analysed.* From UN DIESA, 1982, 36-7.

dwt to transport the ore to shore-based processing plants. Overhaul, maintenance, supply, and storage facilities would be required at the port to accommodate the needs of the ships involved. Ports would also be required to accommodate the slurry pipeline which would transport the ore from the port to the processing plant. The mining ships and transport ships would place high demands on those harbour and docking facilities which presently cannot accommodate a large increase in ship traffic." [Holtz, 1988, 171].

(vi) *Ocean recreational uses* consisting of cruising, fishing, settlements in ocean islands, are increasingly managed by multinational firms so they progress in individual ocean areas only to the extent in which they are included in world-wide strategies.

(vii) *Ocean research* is increasingly managed or co-ordinated by international institutions and decision centres, which—particularly when acting in the context of global change-inspired programmes—tend to focus attention on processes involving the whole ocean and to relate analyses of single ocean areas to comprehensive views of the ocean as a whole. This situation encourages the interaction between corporate strategies and research [Burroughs, 1981 and 1986].

(viii) Ocean *environmental management* is due to the international agreements on pollution from ships, oil and gas installations, etc., providing homogeneous criteria for all ocean areas or, at least, for all areas pertaining to a given category.

(ix) *United Nations action* offers a common background to the management of ocean areas both in terms of resource exploitation and environmental protection and preservation.

Because of these processes inputs from the external environment to ocean areas should not be underestimated, as well as the reaction of these areas to the external environment. As for the criteria by which this issue can be approached, general system theory suggests adopting three principles. First, the natural and social components of the external environment should be considered together, namely by all their inputs to ocean areas and the impulses that they receive from them. This holistic approach is appropriate also for investigating the responses of the ocean area to the inputs from outside. Secondly, relationships between ocean areas and their external environment are not to be regarded in a cause-effect context, i.e. in a context consisting of inputs which, moving from the external environment, deterministically provoke the evolution of the coastal area. Thirdly, because of these indeterministic relations the ocean area, in its interaction with the external environment, can also undergo morphogeneses when it receives small inputs. In other words, in order to set up appropriate input patterns, the ocean area should be regarded as a non-trivial machine.

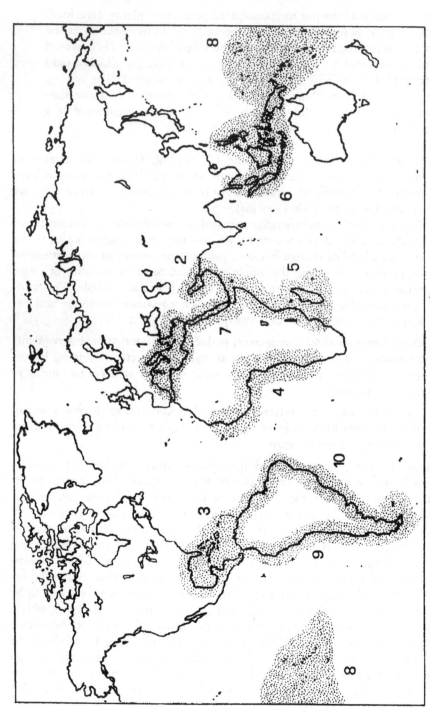

Figure 10.5 *UNEP regional seas programme*. Action plans: (1) Mediterranean region; (2) Kuwait region; (3) Caribbean region; (4) West and Central Africa region; (5) East Asia region; (6) East Africa region; (7) Red Sea and Gulf of Aden region; (8) South-West Pacific region; (9) South-East Pacific region; (10) South-West Atlantic region. From Thacher, 1983, 451.

A latere of these general statements it should be also taken into account that, in many parts of the world, ocean areas receive inputs from neighbouring coastal areas. In these cases coastal areas act as components of the external environment interacting with the bordering ocean area. This is one of the most important features of the present stage of sea management. In fact, as the coastal uses and subsequent environmental impacts grow, involving larger and larger areas of the waters adjacent to the continental margin, inputs are also generated to nearby ocean areas influencing both sea uses, such as navigation and research, and the environment, which could be influenced by a growing demand for chemical and biological oxygen. Until now relationships between ocean and coastal areas have not been intensively investigated in the literature and the coastal areas have never been considered as a part of the external environment of the ocean area. Nevertheless, because of the evolution of these areas it is expected that in the near future this issue will be dealt with.

10.3 THE CONCEPTUAL ISSUE: OCEAN AREA AND OCEAN REGION

It is self-evident that, at least as a first approach, the interest in distinguishing the ocean area from the ocean region appears less interesting than in the coastal area case. Two reasons support this conviction:

(i) the ocean area is not involved in as large a range of sea uses and it is not affected by such a strong human pressure as the coastal area, hence it is characterized by a much lesser degree of organisational complexity than the coastal area;

(ii) the ocean area consists only of the marine environment, so it is involved in much less environmental complexity than the coastal area which, by its nature, is a natural system consisting of the sea and land.

The latter statement is worthy of a discussion focused on the role of islands and archipelagoes in the context of ocean management. When the ocean area is identified through legal criteria, it cannot include terra firma because islands and surrounding marine zones, being subject to national jurisdiction, are regarded as pertaining to the coastal area. In this case also islands and archipelagos located beyond the continental margin and belonging to ocean space are referred to coastal management. As a result, coastal management criteria are applied also to a physical ocean context. On the other hand, if the ocean area is identified through physical criteria, islands and archipelagos, which arose born as a consequence of ocean processes and are located beyond the continental margin, may be regarded as oceanic. So it should be admitted that the ocean area consists also of terra firma, i.e. islands and archipelagos.

As a first approach, this distinction seems to be only theoretical—a mere exercise. Nevertheless, the natural environment and, *optimo iure*, the ecosystem belonging to the marine spaces surrounding ocean islands and archipelagos, *latitudes being equal*, are very different from those surrounding continental islands and archipelagos. As a consequence, management of both resources and the environment should respond to different criteria. On the one hand, the ocean should be managed in its entirety with the same principles, and this calls for international regulations and policies. On the other hand, ocean islands and archipelagos, as well as the surrounding large marine areas, are subject to national status. As a consequence, it is worth wondering whether and to what extent it is possible that national regulations and policies can be everywhere consistent with the action of an international régime and policy which, by its nature, assumes the protection and preservation of the ocean environment as its main goals. There is a potential gap arising from the legal framework: of course, it will not necessarily influence ocean management, if international co-operation is capable of providing appropriate codes of conduct.

At this point, one may wonder whether, also in the ocean space, it is opportune to distinguish the organised area from the region. In short, the question is whether ocean management could acquire such strong and complicated features as to create ocean regions. These following answers can be made:

(i)   As far as ocean areas consisting only of marine spaces are concerned, at the present time the number of natural elements involved in management, as well as those of human activities and facilities, are too small allow us to think that ocean regions exist, or are emerging in the short term.

(ii) As far as ocean areas, including islands and archipelagos are concerned, there is no doubt that in some parts of the world the number of natural elements, as well those of human activities and facilities, are so large as to justify thinking that ocean regions exist. In particular, what induces us to state this is the human pressure that in some parts—e.g. in some Pacific archipelagos—is becoming unexpectedly high.

Of course, from the juridical point of view the areas in ii) are considered as coastal because they belong to national jurisdictional zones but in reality they are oceanic. If this statement is agreed three deductions seem rational:

(i)   the ocean region can occur only where there are oceanic islands and archipelagos;

(ii) supposing that the ocean natural elements range from N_1 to N_n, the ocean region takes shape where the management is pushed beyond a given threshold N_m, involving the sub-range (N_{m+1}, N_n); and

(iii) supposing that the human elements (activities and facilities) range from H_1 to H_n, the ocean management comes to light where the management is pushed beyond the threshold Hm, involving the sub-range (H_{m+1}, H_n).

As can be seen, this reasoning is similar to that developed on the coastal region (Chapter 7).

10.4 LOOKING FOR CONCLUSIONS

Making efforts to clarify goals and methodologies of both coastal and ocean management Vallejo [1988, 213-4] states that these two areas should be managed in an integrated way. This is rational, of course. In order to favour the achievement of this objective it is necessary that both the concepts of coastal and ocean areas, and subsequently the concepts and principles of both coastal and ocean management, are extensively discussed and, particularly, that attention is focused on the two criteria—the legal and the natural—for delimiting them. The consciousness that the employment of legal criteria only is straight forward but not satisfactory should be reinforced, particularly when the willingness to appropriately protect coastal and ocean environments is in the mind of both researchers and managers. This concern, which may be faced by the literature in the 1990s, is not only theoretical because, *inter alia*, it involves the ways in which these two kinds of area are thought of and biases the criteria by which the marine environment is managed.

It should also be clarified how *integration* between coastal and ocean management could be achieved, which implies that the relationships between these two areas be explained. As can be seen, there is a theoretical feed-back relationship: the manner by which the coastal and ocean areas are conceived influences the ways in which reciprocal relationships are explained; explanation influences management patterns; according to the management patterns, the concepts of these areas are formulated and widely agreed. This chain of causation is circular.

An appropriate way to frame these relationships is provided by the general system-based approach, according to which the coastal area is regarded as a main component of the external environment of the ocean area, and vice versa. In fact, between the structure and its external environment a sort of *consensual domain* is established—as Maturana [1978] stated through the autopoiesis concept. This implies that to some extent each structure produces its own external environment and that, in its turn, the external environment acts as a structure. According to the management patterns which it undergoes, each coastal area brings forth its ocean external environment; and each ocean area, according to the international régime and management patterns which it undergoes, determines how the surrounding coastal areas should act as its external environment.

Expressing the hope that ocean management is implemented in the near future Vallejo [*ibid.*, 213-214] thinks that the relationships between coastal and ocean management will follow one of these three paths: i) a "continuation of the present state of affairs", which brings about a growing gap between these two management fields, since coastal management is expected to progress faster than ocean management; ii) "growing activities in CAM and OM undertaken independently"; iii) "an increase in OM (Ocean Management) initiatives that are integrated with CAM (Coastal Area Management) efforts". The third path is desirable, of course, but not necessarily the most likely to take place.

CONCLUSIONS

THE EVOLUTION OF SEA MANAGEMENT

While the tenth anniversary of the Third UN Convention on the Law of the Sea and the twentieth anniversary of the UN Conference on the Human Environment are being celebrated, many managers and scientists are investigating the evolution which both the goals and contents of sea management are undergoing. In these analyses the interaction between national policies and international co-operation, on the one hand, and theoretical and methodological directions provided by present scientific thought, on the other, are acquiring importance.

Nations tend to create the widest possible marine zones within their jurisdictional umbrellas and, in the meanwhile, to establish co-operation in protecting and preserving coastal areas, semi-enclosed and enclosed seas. International co-operation has achieved good results in promoting general and regional conventions to protect and preserve the sea from pollution and hazards but it has not been successful in creating the régime for the exploitation of deep seabeds.

The scientific approach has advanced on all fronts. On one hand, oceanography has been progressing giving impetus also to applied research dealing with resource uses. Because of this it has been justified in regarding itself as the sole research field to be involved in marine affairs without giving great importance to the prospect of setting up ocean science. On the other hand, social sciences—such as the economics of fisheries, the economics and geography of transportation, and geopolitics—have provided notable sectoral views of the sea. As a result, the scientific world has been fragmented, every discipline defending and enlarging its research field independently.

The justification of this fragmented set of scientific approaches has been weakening since the birth and development of coastal management, and later that of ocean management, which requires multi-disciplinary approaches. This demand cannot be met only by the convergence of various disciplines on the same subject (sea management) and on the same area (coastal or oceanic). Sea management needs something much more binding: a scientific integrated approach. To this end the disciplines involved should agree on a unique conceptual framework and a common view of reality: in other words, they are asked to create appropriate isomorphisms. As a consequence, two classic issues—interdisciplinarity and relationships between natural and social sciences—come to the

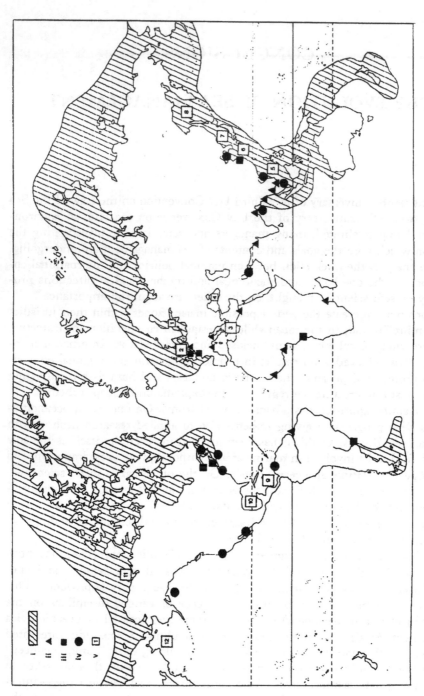

Figure C.1. *Some physical features influencing coastal management*: (i) widest shelves, (ii) major deltas and (iii) estuaries, (iv) bay and gulfs surrounding high urbanized areas and involved in intense maritime traffic, (v) major semienclosed seas: 1. Baltic Sea, 2. North Sea, 3. Mediterranean Sea, 4. Black Sea, 5. South China Sea, 6. East China Sea, 7. Sea of Japan, 8. Sea of Okhtosk, 9. Caribbean Sea, 10. Gulf of Mexico, 11. Beaufort Sea, 12. Bering Sea.

fore acquiring particular pragmatic importance: the scientific world is involved in a turn-round.

In this context attention shifts to those coastal and ocean areas which are influenced by important natural and human factors to the point of being involved in management. This subject could be faced in a tentative way by considering (i) an essential range of natural factors, (ii) some marine resources which have a actual or potential role, (iii) some features which the human presence in coastal areas has created.

As far as the *natural environment* is concerned (Figure C.1), it may be thought that the best ground for developing sea management is found where shelves are wide, there are deltas and estuaries, as well as gulf and bays, provided with opportunities for human settlements and activities. In this view, most semi-enclosed seas appear to have top priority. In the matter of *marine resources* (Figure C.2) since the late Sixties biological resources and gas fields have been increasing importance while manganese nodules are expected to be the most important mineral resource. As a tentative approach, the *human presence* (Figure C.3) may be perceived by considering major urbanized littoral areas and the largest seaports. *A latere*, choke points (straits, canals and oceanic convergence points) are worth taking into account as the main factors influencing navigation and maritime transportation [Alexander and Morgan, 1988].

The contextual analysis of these features and factors allow us to consider some areas, at the present time deeply involved in sea management—as well as others, which could be involved in the short and mid terms, into the 1990s. It should of course be noted that these are only tentative deductions: their role is to contribute to the discussion on the directions and spatial dimensions of sea management.

North Atlantic Ocean. This space, which was the basis for the palaeo- and neo-industrial stages and competes with the North Pacific Ocean in the post-industrial stage, is provided with wide shelves, oil and gas resources (particularly in the European side) and highly developed fisheries and has some of the most economically important semienclosed seas (North Sea, Gulf of Maine). No large deep-sea mineral resources have been found and the largest zooplankton areas are at high latitudes. In comparison with other oceanic regions its importance is particularly due to an unusual concentration of urban areas (the US and Rhine megalopolises), seaports and maritime routes. Although the semi-enclosed seas of this Atlantic space are not covered by UNEP's programmes, they have an important role for the development of coastal area management patterns.

At the present time areas important to sea management extend from the latitudes of North Carolina to the Gulf of Maine in the western Atlantic and from the Strait of Dover to the Baltic Sea in the eastern Atlantic. In the future northward expansion may be expected towards the polar latitudes, where the shelves are wide, zooplankton rich and perhaps oil and gas resources are promising. In this management context the Mediterranean Sea seems to play the role of an appendix of the eastern Atlantic side: this is not due to its level of resource ex-

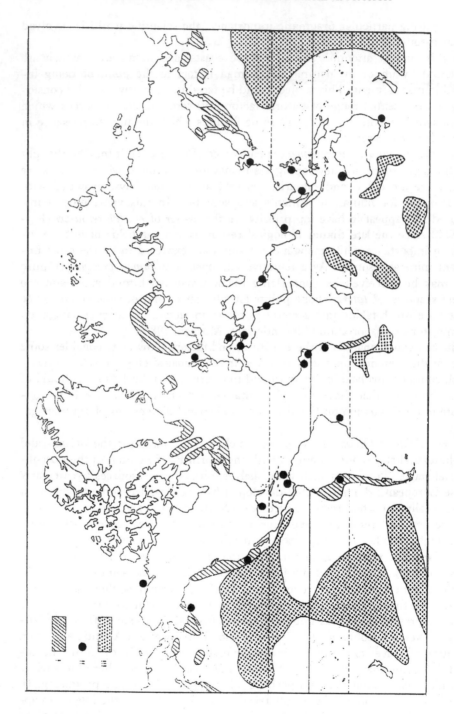

Figure C.2 *Major marine resource distributions*: (i) zooplankton abundance area (more than 500 mg/cu m); (ii) oil and gas production major areas; (iii) manganese nodule deposits.

ploitation—which is much lower than that of the North Sea and adjacent marine spaces—but to the functional links that it has with the Atlantic.

North Pacific Ocean. This region benefits from a unusual context and prospects, particularly on its western side. As a matter of fact, in the temperate belt of the Asian waters, shelves are wide, zooplankton and hydrocarbon fields are abundant, there are the largest number of major seaports in the world and well extended megalopolises. In addition, here is the largest number of semi-enclosed seas, bays and gulfs in the world which, *inter alia*, are covered by UNEP's plans. The eastern side has no similar endowment of gulfs and bays and has no semienclosed seas, but benefits from rich zooplankton and oil and gas fields. From San Francisco Bay to the Gulf of California there extends one of the most important urban zones, including areas endowed with the most advanced technologies. Lastly the existence of rich nodule deposits between the Clarion and Clipperton fractures, which are located inside the US Exclusive Economic Zone, justifies our thinking that the eastern side of the North Pacific will probably be the first area in the world in which complex ocean management will take place.

As far as mineral resources are concerned, management is expected to extend northward, both on the eastern side (Gulf of Alaska and Bering Sea) and in the western side (Sea of Okhotsk). Unfortunately both sides are deeply affected by seismicity and consequent volcanism but, from the management point of view, this circumstance makes this area the most stimulating in the world for the creation of patterns aimed at protection from catastrophism.

The inter-tropical seas: Asia. The marine area extending from the Bay of Bengal to the South China Sea could be regarded as a fascinating southern prolongation of the North Western Pacific area. Shelves are wide, hydrocarbon and biological resources are abundant, the concentration of seaports is considerable, as well as that of choke points, and UNEP's activity is promoting co-operation. As a result, the web of semi-enclosed seas, bays, gulfs and archipelagic areas of these marine spaces is expected to be the most important in the inter-tropical belt and might be the most promising for the implementation of management criteria and methodologies in warm oceanic waters.

The inter-tropical seas: Gulf of Mexico and Caribbean. Shelves are not wide, but oil fields are rich, maritime transportation is well developed and the human pressure in coastal zones is high. In addition, the most important choke point (Panama) for inter-oceanic relationships is located here. These factors, together with the international co-operation encouraged by the UNEP's Action Plan, substantially contribute to make this marine space important for coastal area management.

The inter-tropical seas: Pacific islands and archipelagos. In some of these island areas there is the implementation of recreational uses and biological and mineral resource exploitation, while human pressure is growing. From the legal point of view, the management which is going to be established is coastal, but it involves the ocean environment requiring special criteria and methodologies.

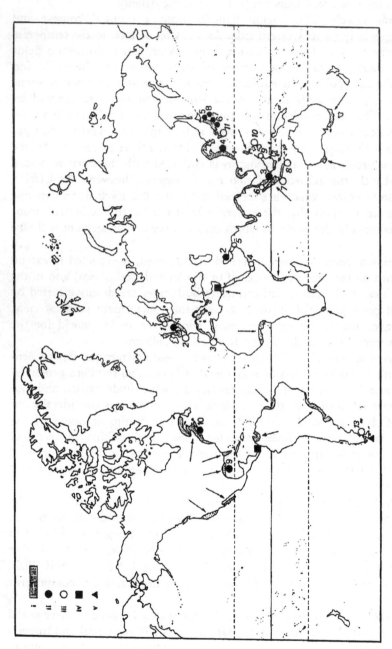

Figure C.3 *Features and factors of the human presence in coastal areas*: (i) *urban belts with high population density*; (ii) *major seaports* (more than 100 million tons per year): 1. Rotterdam, 2. Kharg, 3. Singapore, 4. Shanghai, 5. Chiba, 6. Kobe, 7. Nagoya, 8. Yokohama, 9. New Orleans, 10. New York; (iii) *choke points, straits*: 1. Danish straits, 2. Strait of Dover, 3. Strait of Gibraltar, 4. Bab el Mandeb, 5. Strait of Hormuz, 6. Malacca-Singapore straits, 7. Sunda Strait, 8. Lombok Strait, 9. Balabac Strait, 10. Surigao Strait, 11. Osumi-kaikyo, 12. Bering Strait, 13. Strait of Magellano; (iv) *canals*: 1. Suez, 2. Panama; (v) *oceanic convergence points*: off of the Cape of Good Hope. Choke points are identified according to L.M. Alexander, 1988.

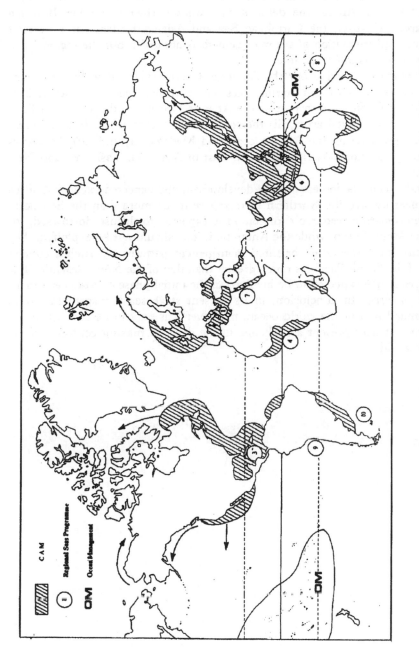

Figure C.4 *Present and potential sea management areas*: (i) coastal area management (CAM) involving wide shelves; (ii) areas covered by the UNEP's Regional Seas Programme, which led to the formulation of Action Plans: 1. Mediterranean, 2. Kuwait, 3. Caribbean, 4. West and Central Africa, 5. East Africa, 6. East Asia, 7. Red Sea and Gulf of Aden, 8. South-West Pacific, 9. South-East Pacific, 10. South-West Atlantic; (iii) potential ocean management (OM).

Other inter-tropical seas. These consist of the Gulf of Guinea and the Atlantic waters of Central Africa, the delta of the Amazon river and other Brazilian coastal areas. *Vis-à-vis* the Caribbean Sea and Asian inter-tropical seas, their role in the implementation of sea management is not great, but the regional and local importance is not negligible.

Southern areas. Leaving out the Antarctic Ocean—which is a different matter—the interest of southern marine areas for sea management is presently limited to (i) the south-western part of the Atlantic Ocean, i.e. the Rio de la Plata estuary and near coastal areas, (ii) the south-western part of the Indian Ocean, i.e. the waters between the African coast and Madagascar, and (iii) the waters surrounding the North-West and South-East of Australia, Tasmania and New Zealand.

As can be seen, the largest present development and expected expansion of sea management involve the Northern Hemisphere much more than the Southern. In addition, marine resource exploitation is expected to diffuse northward, towards the Arctic Ocean, while the Antarctic waters should not be exploited: they should aim for preservation-oriented management patterns. In such a context the role of the North Pacific is much greater than that of the North Atlantic. The Pacific area is also expected to be involved in the initial phase of *real* ocean management patterns. In conclusion, in its present stage sea management covers only a limited part of the world ocean, moreover it is only in its take off phase. Both circumstances present a challenge to the decision making centres and the scientific world.

REFERENCES

Abrahmsson, B.J. (1983). Merchant shipping in transition: an overview. In *Ocean Yearbook 4*, ed. E. Borgese & N. Ginsburg. University Chicago Press, Chicago, pp. 121–39.

Alexander, L.M. (1986). Large Marine ecosystems as regional phenomena. In *Variability and management of large marine ecosystems*, ed. K. Sherman & L. M. Alexander. AAAS Selected Symposium 99, Westeview Press, Boulder, pp. 239–40.

Alexander, L.M. (1989). Large Marine ecosystems as global management units. In *Biomass Yelds and Geography of Large Marine Ecosystems*, ed. K. Sherman & L. M. Alexander. AAAS Selected Symposium 111, Westeview Press, Boulder, pp. 339–44.

Alexander, L.M. & Morgan, J.R. (1988). Choke points of the world ocean: a geographic and military assessment. In *Ocean Yearbook 7*, ed. E. Mann Borgese & N. Ginsburg. University Chicago Press, Chicago, pp. 340–55.

Anderson, A.T. (1982). The Ocean Basins and Ocean Water. In *Ocean Yearbook 3*, ed. E. Mann Borgese & N. Ginsburg. University Chicago Press, Chicago, pp. 135–56.

Anderson, S.H. (1980). The role of recreation in the marine environment. In *Ocean Yearbook 2*, ed. E. Mann Borgese E. & N. Ginsburg. University Chicago Press, Chicago, pp. 183–98.

Archer, J.H. & Knecht, R.W. (1987). The U.S. national coastal zone management programme—problems and opportunities in the next phase. *Coastal Management*, **15**, 2, 103–20.

Armstrong, J.M. & Ryner, P.C. (1981). *Ocean Management. A New Perspective*. Ann Arbor Science, New York.

Attard, D.J. (1987). *The Exclusive Economic Zone in international law*. Clarendon Press, Oxford.

Barnaby, C.F. (1985). Superpower military activities in the world's oceans. In *Ocean Yearbook 5*, ed. E. Mann Borgese & N. Ginsburg. University Chicago Press, Chicago, pp. 223–39.

Bass, G.F. (1980). Marine archaeology: a misunderstood science. In *Ocean Yearbook 2*, ed. E. Mann Borgese & N. Ginsburg. University Chicago Press, Chicago, pp. 137–52.

Bertram, K.M. & Santini, D.J. (1988). United States emergency response capabilities for hazardous materials incidents in U.S. and nearby coastal zones. In *Ocean Yearbook 7*, ed. E. Mann Borgese E., N. Ginsburg & J.R. Morgan. University Chicago Press, Chicago, pp. 159–76.

Bewers, J.M. & Garrett, C.J.R. (1987). Analysis of the issues related to sea dumping of radioactive wastes. *Marine Policy*, **11**, 2, 105–24.

Bird, E.C. (1985). *Coastline Changes. A Global Review*. J. Wiley & Sons, New York.

Birnie, P.W. (1987). Piracy: past, present and future. *Marine Policy*, **11**, 3, 163–83.

Blake, G. (1987). Worldwide maritime boundary delimitation: the state of play. In *Maritime Boundaries and Ocean Resources*, ed. G Blake. Croom Helm, London, pp. 1–14.

209

Blissenbach, E. & Nawab, Z. (1982). Metalliferous sediments of the seabed: the Atlantis II deep deposits of the Red Sea. In *Ocean Yearbook 3*, ed. E. Mann Borgese & N. Ginsburg. University Chicago Press, Chicago, pp. 77–104.

Boaden, P. & Seed, R.D. (1985). *An introduction to coastal ecology*. Black and Son, London.

Boczek, B.A. (1986). The concept of regime and the protection and preservation of the marine environment. In *Ocean Yearbook 6*, ed. E. Mann Borgese & N. Ginsburg. University Chicago Press, Chicago, pp. 271–97.

Braudel, F. (1979). *Civilisation matérielle, économie et capitalisme (XV–XVIII siècle). Les temps du monde*. Colin, Paris.

Briscoe, J. (1988). Islands in Maritime Boundary Delimitation. In *Ocean Yearbook 7*, ed. E. Mann Borgese, N. Ginsburg & J.R. Morgan. University Chicago Press, Chicago, pp. 14–41.

Broadus, J.M. & Gaines, A.G. (1987). Coastal and marine areas management in the Galapagos Islands. *Coastal Management*, **15**, 1, 75–88.

Brown, E.D. (1984). *Sea-bed energy and mineral resources and the law of the sea*, volume I, *The areas within national jurisdiction*. Graham and Trotman, London.

Brown, E.D. & Churchill, R.R. (1985). *The UN Convention on the Law of the Sea: impact and implementation*. The Sea Institute, Honolulu.

Buchholz, H.J. (1987). *Law of the sea zones in the Pacific Ocean*. Institute of Asian Affairs, Hamburg.

Burbridge, P.R., Dankers, N. & Clark, J.R. (1989). Multiple-use assessment for coastal management. In *Coastal Zone '89*, ed. O.T. Magoon & al. American Society of Civil Engineers, New York, Vol. 1, pp.16–32.

Burroughs, H.R. (1981). OCS oil and gas: relationships between resource management and environmental research. *Coastal Zone Management Journal*, **9**, 1, 77–88.

Burroughs, H.R. (1986). Seafloor area within reach of petroleum technology. *Ocean Management*, **10**, 125–35.

Camhis, M. & Coccossis, H. (1982). Coastal planning and management perspectives. *Ekistics*, **49**, 293, 92–7.

Carter, R.W.G. (1987). Man's Response to Sea-level Change. In *Sea Surface Studies. A Global View*, ed. R.J.N. Devoy. Croom Helm, London, pp. 464–98.

Chappell, J. (1987). Ocean Volume Change in the History of Sea Water. In *Sea Surface Studies. A Global View*, ed. R.J.N. Devoy. Croom Helm, London, pp. 33–56.

Charlier, R.H. (1983). Water, energy, and nonliving ocean resources. In *Ocean Yearbook 4*, ed. E. Mann Borgese & N. Ginsburg. University Chicago Press, Chicago, pp. 75–120.

Clark, R.B. (1986). *Marine Pollution*. Clarendon Press, Oxford.

Couper, A.D. (ed.). (1983) *Atlas of the Oceans*. Times Book, London.

Couper, A.D. (1987). Marine resources and environment. *Progress in Human Geography*, **2**, 296–308.

Day, J.C. & Gamble, Don B. (1990). Coastal zone management in British Columbia: an institutional comparison with Washington, Oregon, and California. *Coastal Management*, **18**, 115–41.

Devoy, R.J.N. (1987). Introduction: First Principles and the Scope of Sea-Surface Studies. In *Sea Surface Studies. A Global View*, ed. R.J.N. Devoy. Croom Helm, London, pp. 1–30.

Drodzov, A.V. (1990). Coastal zone in the global environment monitoring system. Paper presented to *Meeting of Commission on Marine Geography*, International Geographical Union, Beijing (August 10–11), proceedings in print.

Earney, F.C.F. (1987). The United States Exclusive Economic Zone: mineral resources. In *Maritime Boundaries and Ocean Resources*, ed. G. Blake. Croom Helm, London, pp. 162–81.

Edwards, S.F. (1989). Estimates of future demographic changes in the coastal zone. *Coastal Management*, **17**, 3, 229–40.

Ford, G., Niblett, C. & Walker, L. (1987). *The future for ocean technology*. Pinter, London.

Forsund, S. & Strom, S. (1987). *Environmental economics and management: pollution and natural resources*. Croom Helm, London.

Frankel, E.G. (1989). Shipping and its role in economic development. *Marine Policy*, **13**, 1, 22–42.

Fricker, A. & Forbes, D.L. (1988). A system of coastal description and classification. *Coastal Management*, **16**, 4, 111–38.

Frosh, R.A., Hollister, C.D. & Deese, D.A. (1978). Radioactive waste disposal in the oceans. In *Ocean Yearbook 1*, ed. E. Mann Borgese & N. Ginsburg. University Chicago Press, Chicago, pp. 340–59.

Geddes, P. (1915). *Cities in evolution, an introduction to the town planning movement and the study of cities*. William & Norgale, London.

GESAMP (IMCO/FAO/UNESCO/WHO/IAEA/UN/UNEP) (1982). *The Health of the Oceans*, UNEP Regional Seas Reports and Studies No. 16. UNEP, Athens.

Goldberg, E. (1985). The ocean as a waste space. In *Ocean Yearbook 5*, ed. E. Mann Borgese & N. Ginsburg. University Chicago Press, Chicago, pp. 150–61.

Hågestrand, T. (1953). *Innovationsforloppet ur korologisk synpunkt*, Lund. English translation by A. Pred, 1968, *Innovation, diffusion as a spatial process*. University Chicago Press, Chicago.

Harrison, P. (1983). Offshore oil and gas: transportation, and coastal impact. In *Ocean Yearbook 4*, ed. E. Mann Borgese & N. Ginsburg. University Chicago Press, Chicago, pp. 319–46.

Harvey, D. (1969). *Explanation in geography*. Arnold, London.

Hayuth, Y. (1988). Changes on the waterfront: a model-based approach. In *Revitalising the waterfront. International dimensions of dockland redevelopment*, ed. B.S. Hoyle, D.A. Pinder & M.S. Husain. Belhaven Press, London, pp. 52–64.

Holt, S. (1978). Marine Fisheries. In *Ocean Yearbook 1*, ed. E. Mann Borgese & N. Ginsburg. University Chicago Press, Chicago, pp. 38–83.

Holtz, K.L. (1988). Implications for coastal managers of hard mineral development in the U.S. Exclusive Economic Zone. *Coastal Management*, **16**, 2, 167–82.

Hoyle, B.S. (1988). Development dynamics at the port-city interface. In *Revitalising the waterfront. International dimensions of dockland redevelopment*, ed. B.S. Hoyle, D.A. Pinder & M. S. Husain. Belhaven Press, London, pp. 3–19.

Hulm, P. (1983). *A strategy for the seas. The regional seas programme past and future*. UNEP, Geneva.

IGBP (1990). *Global Change*, Report No. 12, *The Initial Core Projects*. Stockholm, ICSU.

Interfutures (1979). *Facing the Future. Mastering the Probable and Managing the Unpredictable*. OCSE, Paris.

IOC (1984). *Ocean science for the year 2000*. UNESCO, Paris.

Jacobson, H.K. & Price, M.F. (1990). *A framework for research on the Human Dimensions of Global Environmental Change*. ISSC, Paris.

Jolliffe, I.P. & Patman, C.R. (1985). The coastal zone: the challenge. *Journal of Shoreline Management*, **1**, 3–36.

Johnson, D.C. (1989). Ocean islands and coastal zone management. In *Coastal Zone '89*, ed. O.T. Magoon O.T. & al.. American Society of Civil Engineers, New York, 3, pp. 2161–76.

Johnson, C.J. & Clark, A.L. (1988). Expanding horizons of Pacific minerals. In *Ocean Yearbook 7*, ed. E. Mann Borgese, N. Ginsburg & J.R. Morgan. University Chicago Press, Chicago, pp. 145–58.

Keeble, D.E. (1967). Models of economic development. In *Models in geography*, ed. R.J. Chorley & P. Haggett. Methuen, London, pp. 243–302.

Kenchington R.A. (1990). *Managing Marine Environments*. Taylor and Francis, London.

Kennett, J. (1982). *Marine geology*. Prentice-Hall, London.

Kent, P. (1980). *Minerals from the marine environment*. Arnold, London.

King, A. (1980). *The State of the Planet*, Report of IFIAS. Pergamon Press, Oxford.

Klarin, P. & Hershman, M. (1990). Response of coastal zone management programs to sea level rise in the United States. *Coastal Management*, 18, 143–65.

Kliot, N. (1987). Maritime boundaries in the Mediterranean: aspects of cooperation and dispute. In *Maritime Boundaries and Ocean Resources*, ed. G. Blake. Croom Helm, London, pp. 208–26.

Langford, S.R. (1987). Common fishery resources and maritime boundaries: the case of the Channel Islands. In *Maritime Boundaries and Ocean Resources*, ed. G. Blake. Croom Helm, London, pp. 89–116.

Laszlo, E. (1988). Evolutionary dynamics in nature and society. Steps toward a cross-disciplinary theory. *Revue Internationale de Systémique*, 2, 2, 109–21.

Le Moigne, J.-L. (1984). *La théorie du systéme général. Théorie de la modélisation*. PUF, Paris (first edition, 1977).

Le Moigne, J.-L. (1985). The intelligence of complexity. In *The science and praxis of complexity*. United Nations University Symposium, Montpellier, pp. 35–61.

Leschine, T.M. (1988). Ocean waste disposal management as a problem in decision-making. *Ocean and Shoreline Management*, 11, 1, 5–29.

Levy, J.P. (1988). Towards an integrated marine policy in developing countries. *Marine Policy*, 12, 4, 326–42.

Lucas, K.C. & Loftas, T. (1982). FAO's EEZ program: helping to build the fisheries of the future. In *Ocean Yearbook 3*, ed. E. Mann Borgese & N. Ginsburg. University Chicago Press, Chicago, pp. 38–75.

Mann Borgese, E. (1986). *The future of the oceans*. Harvest House, Montreal.

Margalef, R. (1985). Ecosystems: diversity and connectivity as measurable components of their complication. In *The science and praxis of complexity*. United Nations University Symposium, Montpellier, pp. 228–44.

Maturana, H. R. (1978). Biology of language: the epistemology of reality. In *Psycology and biology of language and thought*, ed. G.A. Miller & E. Leneberg E. Academic Press, New York, pp. 27–64.

Millot, C. (1989). La circulation générale en Méditerranée occidentale. *Annales de Géographie*, 98, 549, 497–515.

Mitchell, J.K. (1982). Coastal Zone Management: A Comparative Analysis of National Programs. In *Ocean Yearbook 3*, ed. E. Mann Borgese & N. Ginsburg. University Chicago Press, Chicago, pp. 258–319.

Mitchell, J.K. (1986). Coastal management since 1980: the U.S. experience and its relevance for other countries. In *Ocean Yearbook 6*, ed. E. Mann Borgese & N. Ginsburg. University Chicago Press, Chicago, pp. 319–45.

Morgan, J.R. (1989a). Marine regions: myth or reality? *Ocean and Shoreline Management*, 12, 1, 1–21.

Morgan J.R. (1989b). Large marine ecosystems in the Pacific Ocean, In *Biomass Yelds and Geography of Large Marine Ecosystems*, ed. K. Sherman & L. M. Alexander, AAAS Selected Symposium 111. Westeview Press, Boulder, pp.377–94.

Morgan, J.R.(1990). Large marine ecosystems of the Pacific Rim. Paper to *The Large Marine Ecosystem (LME) Concept and its application to regional marine resource management*. Monaco, proceedings in print.

Morin, E. (1985). On the definition of complexity. In *The science and praxis of complexity*. United Nations University Symposium, Montpellier, pp. 62–8.

Mörner, N.-A. (1987). Models of global sea-level changes. In *Sea-level changes*, ed. M.J.

Tooley & I. Shennan. Blackwell, Oxford, pp. 332–355.

Morris, M.A. (1982). Military aspects of the Exclusive Economic Zone. In *Ocean Yearbook 3*, ed. E. Mann Borgese & N. Ginsburg. University Chicago Press, Chicago, pp. 320–48.

Mukang, H., Shusong, Z. & Luiqing, G. (1988). China's coastal environment, utilization and management. In *Ocean Yearbook 7*, ed. E. Mann Borgese, N. Ginsburg & J.R. Morgan. University Chicago Press, Chicago, pp. 223–40.

Mumford, L. (1934). *Technics and Civilization*. New York.

Odell, P.R. (1978). Oil and gas exploration and exploitation in the North Sea. In *Ocean Yearbook 1*, ed. E. Mann Borgese & N. Ginsburg. University Chicago Press, Chicago, pp. 139–59.

Odum, E.P. (1971). *Fundamentals of ecology*. Saunders, Phidaldelphia.

Orford, J. (1987). Coastal Processes: The Coastal Response to Sea-level Variation. In *Sea Surface Studies. A Global View*, ed R.J.N. Devoy. Croom Helm, London, pp. 415–63.

Pardo, A. (1978). The Evolving law of the Sea: A Critique of the Informal Composite Negotiating Text (1977). In *Ocean Yearbook 1*, ed. E. Mann Borgese & N. Ginsburg. University Chicago Press, Chicago, pp. 9–37.

Pirazzoli, P.A. (1987). Sea-level changes in the Mediterranean. In *Sea-level changes*, ed. M.J. Tooley & I. Shennan. Blackwell, Oxford, pp. 152–81.

Ploman, E.W. (1985). Introduction. In *The science and praxis of complexity*. United Nations University, Montpellier, pp. 7–22.

Post, A.M. (1983). *Deepsea Mining and the Law of the Sea*. Martinus Nijhoff, The Ague.

Prescott, J.R.V. (1985). *The Maritime Political Boundaries of the World*. Methuen, London.

Prescott, J.R.V. (1987a). *Maritime jurisdiction in East Asian seas*. Occasional Papers of the East-West Environment and Policy Institute, No. 4, Honolulu, East–West Center.

Prescott, J.V.R. (1987b). *Straight and archipelagic baselines*, in Maritime Boundaries and Ocean Resources, ed. G. Blake. Croom Helm, London, pp. 38–51, 1987.

Prescott, J.V.R. (1990). The role of national political factors in the management of LMEs: evidence from West Africa. Paper to *The Large Marine Ecosystem (LME) Concept and its application to regional marine resource management*. Monaco, proceedings in print.

Pyeatt Menefee, S. (1988). Maritime terror in Europe and the Mediterranean. *Marine Policy*, **12**, 2, 143–52.

Raftopoulos, E.G. (1988). *The Mediterranean Action Plan in a functional perspective: a quest for law and policy*, MAP (Mediterranean Action Plan) Technical Reports Series No. 25. UNEP, Athens.

Rostow, W.W. (1960). *The Stages of Economic Growth*. University Press, Cambridge.

Rothschild, B. (1983). *Global fisheries: perspectives for the 1980s*. Springer-Verlag, New York.

Santis Arenas, H. (1988). Frontiers and maritime boundaries in the context of territoriality: the Chilean case. In *The new frontiers of marine geography*, ed. H.D. Smith & A. Vigarié. International Geographical Union, Commission on Marine Geography, Rome, pp. 44–71.

Smith, H.D. (1987). Maritime boundaries and the emerging regional bases of world ocean management. In *Maritime Boundaries and Ocean Resources*, ed. G. Blake. Croom Helm, London, pp. 77 88.

Smith, H.D. (1988). The theory and practice of sea use management. In *The new frontiers of marine geography*, ed. H.D. Smith & A. Vigarié. International Geographical Union, Commission on Marine Geography, Rome, pp. 16–32.

Smith, H.D. & Lalwani, C.S. (1984). *The North Sea: sea use management and planning*. University of Wales, Institute of Science and Technology, Cardiff.

Smith, R.W. (1986). *Exclusive Economic Zone Claims. An analysis and primary documents*. Martinus Nijhoff, Dordrecht.

Spence, N.A. & Taylor, P.J. (1970). Quantitative methods in regional taxonomy. *Progress in Geography*, **2**, 3–64.

Stewart, J.R. (1970). Marine Ecumenopolis. *Ekistics*, **29**, 175, 399–418.

Symonds, E. (1978). Offshore oil and gas. In *Ocean Yearbook 1*, ed. E. Mann Borgese & N. Ginsburg. University Chicago Press, Chicago, pp. 114–38.

Thacher, P.S. (1983). UNEP: an update on the regional seas programme. In *Ocean Yearbook 4*, ed. E. Mann Borgese & N. Ginsburg. University Chicago Press, Chicago, pp. 450–61.

Tooley, M.J. (1987). Sea-level studies. In *Sea-level changes*, ed. M.J. Tooley & I. Shennan. Blackwell, Oxford, pp. 1–24

Tracher, P.S. & Meith-Avcin N. (1978). The oceans: health and prognosis. In *Ocean Yearbook 1*, ed. E. Mann Borgese & N. Ginsburg. University Chicago Press, Chicago, pp. 293–339.

Turekian, K.K. (1976). *Oceans*. Prentice-Hall, Englewood-Cliffs.

UNEP (1982). *Achievements and planned development of UNEP's Regional Seas Programme and comparable programmes sponsored by other bodies*, UNEP Regional Seas Reports and Studies, No. 1. UNEP, Nairobi.

UNEP (1989). *State of the Mediterranean Marine Environment*, MAP (Mediterranean Action Plan) Technical Report No. 28. UNEP, Athens.

UNEP (1990). *UNEP Oceans Programme: Compendium of Projects*, UNEP Regional Seas Reports and Studies, No. 19, Rev. 5. UNEP, Nairobi.

United Nations, DIESA (1982). *Coastal Area Management and Development*. Pergamon Press, Oxford.

United Nations, DIESA (1984). *A Guide to ocean thermal energy conversion for developing countries*. UN, New York.

United Nations, OETB (1982). *Assessment of manganese nodule resources*. Graham and Trotman, London.

United Nations, OETB (1984). *Analysis of exploration and mining technology for manganese*. Graham and Trotman, London.

United Nations, Secretary-General (1978). *Progress report of the United Nations Secretary-General*, document E/5971, UN, New York, 1977; reproduced in Ocean Yearbook 1, 1978, pp. 350–75.

Uthoff, D. (1983). Konfliktfeld Nordsee. *Geographische Rundschau*, 6, 282-91.

Vallega, A. (1980). *Per una geografia del mare. Trasporti marittimi e rivoluzioni economiche*. Mursia, Milano.

Vallega, A. (1988). A human geographical approach to semienclosed seas: the Mediterranean case. In *Ocean Yearbook 7*, ed. E. Mann Borgese, N. Ginsburg and J.R. Morgan. University Chicago Press, Chicago, pp. 272–93.

Vallega, A. (1990a). *Ocean change in global change. Introductory geographical analysis*. International Geographical Union, Commission on Marine Geography, Genoa.

Vallega, A. (1990b). *Esistenza, società, ecosistema. Pensiero geografico e questione ambientale*. Mursia, Milano.

Vallejo, M.S. (1988). Development and Management of Coastal and Marine Areas: An International Perspective. In *Ocean Yearbook 7*, ed. E. Mann Borgese & N. Ginsburg. University Chicago Press, Chicago, pp. 205–22.

Van Dyke, J.M. (1988). Ocean disposal of nuclear waste. *Marine Policy*, **12**, 2, 82–95.

Vigarié, A. (1981). Maritime Industrial Development Areas: Structural Evolution and Implication for Regional Development. In *Cityport Industrialization and Regional Development. Spatial Analysis and Planning Strategies*, ed. B.S. Hoyle & D.A. Pinder. Permagon Press, Oxford, pp. 23–37.

Von Bertalanffy, L. (1968). *General System Theory*. Braziller, New York.

Von Foester, H. (1985). Cibernetica ed epistemologia: storia e prospettive. In *La sfida della complessità*, ed. G. Bocchi G. & M. Ceruti. Feltrinelli, Milano, pp. 112–40.

Westing, A.H. (1978). Military impact on ocean ecology. In *Ocean Yearbook 1*, ed. E. Mann Borgese & N. Ginsburg. University Chicago Press, Chicago, pp. 436–66.

Wise, M. (1987). European, national and regional concepts of fishing limits in the European Community. In *Maritime Boundaries and Ocean Resources*, ed. G. Blake. Croom Helm, London, pp. 117–32.

Whittlesey, D. (1954). The regional concept and the regional method. In *American Geography: inventory and prospect*, ed. P.E. James & C.F. Jones. Association of American Geographers, Syracuse University Press, Syracuse, pp. 21–68.

Zeleny, M. (1985). Spontaneous social orders. In *The science and praxis of complexity*. United Nations University Symposium, Montpellier, pp. 312–28.

APPENDIX A

THE SEA USE FRAMEWORK

The concept of sea use is only apparently easy to formulate and to apply in the investigations of coastal and ocean use structures. To begin with it is self-evident that this concept embraces two other concepts: facilities and social behaviour. Facilities are concerned in the investigation because plants, installations, man-made structures—such as seaports, pipelines, offshore platforms and pipelines—are considered. Social behaviour is concerned when activities, such as fishing, transportation and the simple human presence, such as settlements, are considered. If the concept of sea use could be related only to facilities or only to behaviour, the framework of sea uses—as well as the models of use-use and use-environment relationships—-would be very different from the models that have been discussed in this Chapter. In any case, it is very difficult for a model, if it is consistent only with one of these criteria (facilities or activities), to be really useful for managerial purposes.

When a sea use framework includes both facilities and activities (i) some sectors of marine resource exploitation are considered only through the kinds of activities of which they consist, and (ii) other sectors are considered through the kinds of facilities that they employ. Including activities or facilities in the sea use framework depends on the role that the framework has: in some cases it is better to mention activities; in other cases it is more useful to mention the man-made structures located at sea. For instance, in the following sea use framework recreational uses in offshore areas (category 13.2) are considered through activities—such as fishing, sailing, cruising, etc.—while the exploration, exploitation and storage of oil and gas (category 9) is considered through plants and installations.

Nevertheless the real methodological problem does not consist of opting for a framework based only on facilities or only on activities or—as has been done here—contextually on both: in fact all components (facilities and activities) lead to considering *functions of man's presence on the sea*. The real problem is to identify the objectives for which a function—expressed by a facility or by an activity—has been set up. Initially, at the present time these objectives are:

(i) the implementation of progress in the knowledge of the sea;
(ii) the advance in the exploitation of non-renewable resources;
(iii) the implementation and diffusion of renewable resource exploitation, particularly if these act as substitutes for non-renewable resources;
(iv) the progress in the management of the marine environment.

According to the view supporting both the policy of the United Nations and numerous governmental and inter-governmental organizations, as well as the scientific world, sea uses should be clustered and investigated in such a way as to make it understandable:

(i)   whether and how the exploitation of non-renewable resources is being rational-
       ised;

(ii)   whether and how the exploitation of renewable resources is advancing with the
       aim of minimizing the exploitation of non-renewable ones;

(iii) to what extent management tends to minimize environmental impacts and, in
       particular, to avoid man-induced morphogeneses in marine ecosystems.

Bearing these considerations in mind the *sea use framework* is hereunder presented. It
is the introductory step in the setting up of the *sea use-use relationship model*. As can be
seen, in this framework coastal uses have been considered *sensu stricto:* only facilities and
activities that, by their nature, are functionally tied to the sea have been included in the
framework. Frameworks specifically concerned with coastal and ocean structures will be
presented, as appendices, in Chapters 7 and 9.

SEA USE FRAMEWORK

category	**1 SEAPORTS**
sub-categories	*uses*
1.1 waterfront commercial structures	1.1.1 liquid bulk terminals 1.1.2 solid bulk terminals 1.1.3 multiple bulk terminals 1.1.4 general cargo terminals 1.1.5 lo-lo container terminals 1.1.6 ro-ro terminals 1.1.7 lo-lo and ro-ro terminals 1.1.8 heavy and large cargo terminals
1.2 offshore commercial structures	1.2.1 Catenary Anchor Leg Moorings (CALMs) 1.2.2 Exposed Location Single Buoy Moorings (ELSBMs) 1.2.3 Single Buoy Moorings (SBMs) 1.2.4 Single Anchor Leg Moorings (SALMs) 1.2.5 Articulated Loading Platforms (ALPs) 1.2.6 Mooring Towers 1.2.7 Single Buoy Storages (SBSs) 1.2.8 Single Anchor Leg Storages (SALSs) 1.2.9 floating transhipment docks 1.2.10 artificial islands
1.3 dockyards	1.3.1 ship building dockyards 1.3.2 ship repairing dockyards 1.3.3 oil and gas platform building dockyards 1.3.4 oil and gas platform maintenance dockyards
1.4 passenger facilities	1.4.1 cruise ship terminals 1.4.2 ro-ro carrier terminals 1.4.3 hydrofoil terminals

1.5 naval facilities	1.5.1 naval dockyards
	1.5.2 surface vessel terminals
	1.5.3 submarine terminals
	1.5.4 weapon storages
	1.5.5 nuclear material deposits
1.6 fishing facilities	1.6.1 vessel terminals
	1.6.2 fishery facilities
1.7 recreational facilities	1.7.1 sailing vessel terminals
	1.7.2 engine-propelled vessel terminals
	1.7.3 wind surfing facilities
	1.7.4 yacht racing facilities
	1.7.5 semi-submersible and submarine vessel facilities

category	**2 SHIPPING, CARRIERS**
sub-categories	*uses*
2.1 bulk vessels	2.1.1 Ultra Large Crude Carriers (ULCCs)
	2.1.2 Very Large Crude Carriers (VLCCs)
	2.1.3 Medium Size Crude Carriers (MSCCs)
	2.1.4 oil product carriers
	2.1.5 chemical product carriers
	2.1.6 liquefied gas carriers (LGCs)
	2.1.7 liquefied natural gas carriers (LNGs)
	2.1.8 solid bulk carriers
2.2 general cargo vessels	2.2.1 conventional break bulk carriers
2.3 unitized cargo vessels	2.3.1 fully containerships
	2.3.2 semi-containerships
	2.3.3 ro-ro ships
	2.3.4 ro-ro cellular ships
	2.3.5 barge carriers
2.4 heavy and large cargo vessels	2.4.1 special ro-ro carriers
	2.4.2 float on-float off carriers
	2.4.3 sliding on-sliding off carriers
	2.4.4 jacking up-jacking down carriers
	2.4.5 semi-submersible carriers
	2.4.6 submersible carriers
2.5 passenger vessels	2.5.1 ro-ro carriers
	2.5.2 cruise vessels
	2.5.3 conference vessels
2.6 multipurpose vessels	2.6.1 solid bulk combined carriers
	2.6.2 solid and liquid bulk carriers
	2.6.3 bulk and general cargo carriers
	2.6.4 bulk container carriers
	2.6.5 bulk ro-ro carriers
	2.6.6 ro-ro cargo and passenger carriers

category	**3 SHIPPING, ROUTES**
sub-categories	*uses*

3.1 routes	3.1.1 deepsea routes
	3.1.2 mid-sea routes
	3.1.3 short-sea routes
	3.1.4 feeder routes
	3.1.5 mini-landbridges based routes
	3.1.6 landbridges based routes
3.2 passages	3.2.1 straits
	3.2.2 canals
3.3 separation lanes	

category	**4 SHIPPING, NAVIGATION AIDS**
sub-categories	*uses*

4.1 buoy systems	4.1.1 simple buoys
	4.1.2 safe facilities-provided buoys
4.2 lighthouses	4.2.1 simple lighthouses
	4.2.2 radiobeacon-provided lighthouses
4.3 hyperbolic systems	4.3.1 Decca Navigator system
	4.3.2 Short Range Navigation system (SHORAN)
	4.3.3 Long Range Navigation (LORAN) system
	4.3.4 Omega system
	4.3.5 Automatic Radar Plotting Aids (ARPA) system
4.4 satellite systems	4.4.1 Transit system
	4.4.2 Global Positioning system
4.5 inertial systems	4.5.1 Ship's Inertial Navigation System (SINS)

category	**5 SEA PIPELINES**
sub-categories	*uses*

5.1 slurry pipelines	5.1.1 fine coal slurry pipelines
	5.1.2 coarse coal slurry pipelines
	5.1.3 limestone slurry pipelines
	5.1.4 phosphate slurry pipelines
	5.1.5 ore slurry pipelines
	5.1.6 copper slurry pipelines

5.2 liquid bulk pipelines	5.2.1 short submarine oil pipelines
	5.2.2 long submarine oil pipelines
	5.2.3 oil product pipelines
	5.2.4 liquefied coal pipelines
5.3 gas pipelines	5.3.1 natural gas pipelines
	5.3.2 extra natural gas pipelines
	5.3.3 coal gas pipelines
5.4 water pipelines	5.4.1 short distance aqueducts
5.5 waste disposal pipelines	5.5.1 short distance pipelines

category	**6 CABLES**
sub-categories	*uses*
6.1 electric power cables	6.1.1 low voltage cables
	6.1.2 high voltage cables
6.2 telephone cables	6.2.1 telephone cables
	6.2.2 facsimile system provided cables
	6.2.3 data transmission system-tied cables

category	**7 AIR TRANSPORTATION**
sub-categories facilities for	*uses*
7.1 airports	7.1.1 waterfront-located airports
	7.1.2 offshore airports
7.2 others	7.2.1 waterfront-located heliports
	7.2.2 offshore heliports
	7.2.3 hoverports

category	**8 BIOLOGICAL RESOURCES**
sub-categories: facilities for	*uses*
8.1 fishing	8.1.1 pelagic fishing
	8.1.2 demersal fishing
	8.1.3 shellfish
8.2 gathering	8.2.1 seaweed gathering

3.3 farming   8.3.1 fish
                                    8.3.2 oysters
                                    8.3.3 mussels
                                    8.3.4 crustaceous
                                    8.3.5 seaweed

8.4 extra food products   8.4.1 marine medicine

---

category	**9 HYDROCARBONS**

sub-categories: phases	*uses*

9.1 exploration   9.1.1 jack-up rigs
                       9.1.2 semi-submersible platforms
                       9.1.3 drill ships
                       9.1.4 seabed wells
                       9.1.5 suspended wells

9.2 exploitation   9.2.1 extended onshore drills
                       9.2.2 steel framed structures (1)
                       9.2.3 steel platforms (2)
                       9.2.4 concrete platforms (2)
                       9.2.5 hyperbuoyant leg platforms (3)
                       9.2.6 undersea manifolds (3)

9.3 storage   9.3.1 waterfront facilities
                       9.3.2 floating storages
                       9.3.3 submerged storages

(1) up to 50 m depth
(2) up to the outer edge of the continental shelf
(3) beyond the outer edge of the continental shelf

---

category	**10 METALLIFEROUS RESOURCES**

sub-categories: exploitation of	*uses*

10.1 sand and gravel   10.1.1 estuarine dredging
                               10.1.2 seabed dredging
                               10.1.3 channel dredging
                               10.1.4 maintenance dredging

10.2 water column minerals   10.2.1 salt
                               10.2.2 bromine
                               10.2.3 magnesium
                               10.2.4 freshwater production

10.3 seabed deposits	10.3.1 phosphorite nodules
	10.3.2 manganese nodules
	10.3.3 manganese pavement nodules
	10.3.4 muds

category	**11 RENEWABLE ENERGY SOURCES**
sub-categories: environments	*uses*

11.1 wind	11.1.1 wind energy
11.2 water properties	11.2.1 thermal gradient energy
	11.2.2 salinity gradient energy
	11.2.3 density gradient energy
	11.2.4 biomass energy
	11.2.5 hydrogenon production
11.3 water dynamics	11.3.1 wave power
	11.3.2 tidal power
	11.3.3 current power
11.4 subsoil	11.4.1 undersea coalmining

category	**12 DEFENCE**
sub-categories: activity areas	*uses*

12.1 exercise areas	12.1.1 air firing exercise areas
	12.1.2 surface firing exercise areas
	12.1.3 surface bombing exercise areas
	12.1.4 submarine exercise areas
12.2 nuclear test areas	
12.3 minefields	
12.4 explosive weapon areas	12.4.1 mines
	12.4.2 torpedoes
	12.4.3 depth charges
	12.4.4 missiles

category	**13 RECREATION**
sub-categories: areas	*uses*

13.1 onshore and waterfront	13.1.1 sunbathing
	13.1.2 swimming pools

| | 13.1.3 | aquatic sports |
| | 13.1.4 | aquariums |

13.2 offshore

	13.2.1	sunbathing
	13.2.2	swimming, diving
	13.2.3	wind surfing
	13.2.4	snorkelling
	13.2.5	fishing
	13.2.6	sailing
	13.2.7	yacht racing
	13.2.8	yacht cruising
	13.2.9	housing
	13.2.10	exploration

---

category   **14 WATERFRONT MAN-MADE STRUCTURES**

---

sub-categories: areas   *uses*

14.1 onshore and waterfront | 14.1.1 | artificial cliffs |
14.1.2	protection walls
14.1.3	drainage basins
14.1.4	polders
14.1.5	artificial beaches
14.1.6	estuarine reclamation

14.2 offshore

14.2.1	groyne fields
14.2.2	marina breakwaters
14.2.3	artificial reefs
14.2.4	artificial tidal pools
14.2.5	sea-walls
14.2.6	jetties

---

category   **15 WASTE DISPOSAL**

---

sub-categories: sources   *uses*

15.1 urban and industrial | 15.1.1 | incineration systems |
plants | 15.1.2 | treated solid waste disposal |
15.1.3	untreated solid waste disposal
15.1.4	treated liquid waste disposal
15.1.5	untreated liquid waste disposal
15.1.6	nuclear waste disposal

15.2 watercourses

15.2.1	rivers
15.2.2	canals
15.2.3	underground watercourses
15.2.4	run-off

15.3 offshore oil and gas installations	15.3.1 harmful and noxious substance discharges 15.3.2 oil and oily mixture discharges 15.3.3 drilling fluid discharges 15.3.4 sewage discharges 15.3.5 garbage discharges
15.4 dumping	15.4.1 aircraft dumping 15.4.2 ship dumping
15.5 navigation	15.5.1 harmful and noxious substance discharges 15.5.2 sewage discharges 15.5.3 garbage discharges

category	**16 RESEARCH**
sub-categories: contents	*uses*

16.1 water column	16.1.1 physics 16.1.2 chemistry 16.1.3 dynamics 16.1.4 anomalous phenomena
16.2 seabed and subsoil	16.2.1 plate tectonic 16.2.2 geology 16.2.3 geomorphology 16.2.4 volcanism 16.2.5 seismicity 16.2.6 mineralisation processes
16.3 ecosystems	16.3.1 shoreline ecosystems 16.3.2 pelagic ecosystems 16.3.3 demersal ecosystems 16.3.4 food webs
16.4 external environment interaction	16.4.1 hydrological cycle 16.4.2 sedimentary cycle 16.4.3 erosion cycle 16.4.4 heavy metal cycles 16.4.5 organic substance cycles
16.5 special areas and particularly sensitive areas	16.5.1 estuaries 16.5.2 coral reefs 16.5.3 lagoons 16.5.4 Arctic seas 16.5.5 Antarctic seas
16.6 sea use management	16.6.1 jurisdictional zones 16.6.2 single use management 16.6.3 multiple use management 16.6.4 environmental management 16.6.5 comprehensive management

category   **17 ARCHAEOLOGY**

sub-categories: areas   *uses*

17.1 onshore and waterfront 17.1.1 harbours
                              17.1.2 settlements
                              17.1.3 hulls
                              17.1.4 ragments of cargoes
                              17.1.5 fittings

17.2 offshore   17.2.1 submerged harbours
                              17.2.2 submerged settlements
                              17.2.3 hulls
                              17.2.4 weapons
                              17.2.5 dispersed cargoes
                              17.2.6 dispersed fittings

category   **18 ENVIRONMENTAL PROTECTION AND PRESERVATION**

sub-categories: areas   *uses*

18.1 onshore and   18.1.1 wetland conservation
      waterfront   18.1.2 dune conservation
                        18.1.3 nature reserve
                        18.1.4 nature parks
                        18.1.5 special areas and particularly sensitive areas
                        18.1.6 ecosystem protection and preservation
                        18.1.7 species conservation

18.2 offshore   18.2.1 marine reserves
                        18.2.2 marine parks
                        18.2.3 ecosystem protection and preservation
                        18.2.4 species protection and preservation

NOTES
onshore: the submerged area extending landward from the coastline;
waterfront: the interface extending between the sea and terra firma;
offshore: the marine area extending seawards from the coastline.

REFERENCES

Couper, A.D. (ed.) (1983). *The Times Atlas of the Oceans*. Times Books, London,
Ford, G., Niblett, C. & Walker, L. (1987). *The future for Ocean Technology*. Pinter, London.
Hydrographer of the Navy (1985). *The Mariner's Handbook*. London.
Intergovernmental Oceanographic Commission, (1984). *Ocean Science for the year 2000*. Unesco, Paris.

Jolliffe, I.P. & Patman, C.R. (1985). The coastal Zone: the challenge. *Journal of Shoreline Management*, 1, 3-36.

Kent, P. (1980). *Minerals from the marine environment*. Arnold, London.

Rothschild, B.(1983). *Global fisheries: perspective for the 1980s*. Springer-Verlag, New York.

UNEP (1982). *Convention for the Protection of the Mediterranean Sea against Pollution and its related protocols*. UN, New York.

Vallega, A.(1990). *Ocean change in global change. Introductory geographical analysis*. IGU Commission on Marine Geography and University of Genoa, Genoa.

POST SCRIPTUM

THE IDENTIFICATION OF CATEGORIES OF USES

There are two main methods for identifying the categories of sea uses. *First*, the category is thought of as a container including every kind of facilities having a given function, apart from the economic cycle to which their functions belong. For instance, since the main function of the seaport is to provide accessibility to sea carriers, every kind of seaport—merchant, naval, fishing and recreational—is included in the same category: the category of "seaports". *Secondly*, the category includes all facilities and activities belonging to a given cycle. For instance, all elements belonging to fishing and fisheries are assembled in the same category: fishing terminals, vessels, coastal fisheries, and so forth. Here the former option has been accepted because it seems more practical than the latter—although it would be stimulating to set up a framework according to the "cycle-based principle".

Bearing in mind that uses have been classified according to the main goals of sea use management (see Paragraph 5.2) sub-categories have been identified as appropriately as possible using this criterion. As a result, sub-categories are concerned with the nature of facilities, the nature of activities, the areas where facilities and activities are usually located, the nature of resources, the content of the knowledge of the marine environment, and so forth. This occurs also in other clusters of sea uses provided by the literature. To make this approach explicit, the criterion pertaining to the identification of each sub-category is mentioned in the table.

APPENDIX B

SEA MANAGEMENT PATTERNS

The set of sea management patterns, listed hereunder, arises from the basic statements formulated in paragraph 3 of this chapter. This approach has been suggested by Prescott's table [1985, 97] on the geographical classification of claims to the continental margin (Chapter 4, Table 4.4). The interplay between (i) national jurisdictional zones, the (ii) extent and (iii) content of management is taken into account.

Jurisdictional zones. The territorial sea, the continental shelf and the Exclusive Economic Zone are considered. The territorial sea is supposed as extending 12 nm from the baselines. The delimitation of the continental shelf is taken into account only in accordance with the possibility that it covers the continental margin.

Management extent. According to the background supporting the present analysis, coastal and ocean managements are distinguished. The coastal area is supposed as extending up to the outer edge of the continental margin. This is thought of as the natural line landward delimiting an area where the rationale in management can be maximised.

Content of management. In order to identify it, it is assumed that:

(i)   coastal management involves the coastal zone, from the land to the outer edge of the rise;

(ii)   ocean management involves the sea beyond the outer edge of the rise

(iii) comprehensive management involves the marine environment as a whole, from the surface to the subsoil;

(iv)   partial management involves only the seabed and subsoil.

Pattern 1
CONTINENTAL MARGIN NARROWER THAN 200 NM FROM THE BASELINES

LEGAL DATA

(i)   Territorial sea extends up to 12 nm from the baselines;

(ii)   Continental shelf has been delimited;

(iii) Exclusive Economic Zone has not been claimed or agreed.

MANAGEMENT AREAS
1 Up to 12 nm
— coastal management
— comprehensive management

2 From 12 nm to the outer edge of the rise
— coastal management
— partial management

3 From the outer edge of the rise to 200 nm
— ocean management
— partial management

PRACTICABLE MANAGEMENT PATTERNS
1   Up to 12 nm: coastal region
2   From 12 nm to the outer edge of the rise: coastal organized area
3   From the outer edge of the rise to 200 nm: ocean organized area

---

Pattern 2
CONTINENTAL MARGIN NARROWER THAN 200 NM FROM THE BASELINES

LEGAL DATA
(i)   Territorial sea extends up to 12 nm from the baselines;
(ii)   Exclusive Economic Zone has been claimed or agreed.

MANAGEMENT AREAS
1 Up to the outer edge of the rise
— coastal management
— comprehensive management

2 From the outer edge of the rise to 200 nm
— ocean management
— comprehensive management

PRACTICABLE MANAGEMENT PATTERNS
1   Up to the outer edge or the rise: coastal region
2   From the outer edge of the rise to 200 nm: ocean region

Pattern 3
CONTINENTAL MARGIN EXTENDED UP TO 200 NM FROM THE BASELINES

Border-line event

LEGAL DATA
(i)   Territorial sea extends up to 12 nm from the baselines;
(i)   Continental shelf has been delimited;
(iii) Exclusive Economic Zone has not been claimed or agreed.

MANAGEMENT AREAS
1 Up to 12 nm
— coastal management
— comprehensive management

2 From 12 nm to 200 nm
— coastal management
— partial management

PRACTICABLE MANAGEMENT PATTERNS
1   Up to 12 nm: coastal region
2   From 12 nm to 200 nm: coastal organized area

Pattern 4
CONTINENTAL MARGIN EXTENDED UP TO 200 NM FROM THE BASELINES

Border-line event

LEGAL DATA
  (i)   Territorial sea extends up to 12 nm from the baselines;
  (ii) Exclusive Economic Zone has been claimed or agreed.

MANAGEMENT AREAS
1 Up to 200 nm
— coastal management
— comprehensive management

PRACTICABLE MANAGEMENT PATTERNS
1   Up to 200 nm: coastal region

Pattern 5
CONTINENTAL MARGIN WIDER THAN 200 NM FROM THE BASELINES

LEGAL DATA
  (i)   Territorial sea extends up to 12 nm from the baselines;
  (ii)   Exclusive Economic Zone has not been claimed or agreed.
  (iii) The coastal state is able to opt for a criterion to delimit the continental shelf in
         such a way as to include the continental margin within it.

MANAGEMENT AREAS
1 Up to 12 nm
— coastal management
— comprehensive management

2 From 12 nm to the outer edge of the rise
— coastal management
— partial management

PRACTICABLE MANAGEMENT PATTERNS
1   Up to 12 nm: coastal region
2   From 12 nm to the outer edge of the rise: coastal organized area

Pattern 6
CONTINENTAL MARGIN WIDER THAN 200 NM FROM THE BASELINES

LEGAL DATA
(i)   Territorial sea extends up to 12 nm from the baselines;
(ii) Exclusive Economic Zone has not been claimed or agreed.
(iii) The coastal state is not able to opt for a criterion to delimit the continental shelf in such a way as to include the continental margin within it.

MANAGEMENT AREAS
1 Up to 12 nm
— coastal management
— comprehensive management

2 From 12 nm to the seaward limit of the continental shelf
— coastal management
— partial management

The state is not able to manage the continental margin as a whole.

PRACTICABLE MANAGEMENT PATTERNS
1   Up to 12 nm: coastal region
2   From 12 nm to the seaward limit of the continental shelf: coastal organized area

---

Pattern 7
CONTINENTAL MARGIN WIDER THAN 200 NM FROM THE BASELINES

LEGAL DATA
(i)   Territorial sea extends up to 12 nm from the baselines;
(ii) Exclusive Economic Zone has been claimed or agreed.
(iii) The coastal state is able to opt for a criterion to delimit the continental shelf in such a way as to include the continental margin within it.

MANAGEMENT AREAS
1 Up to 200 nm
— coastal management
— comprehensive management

2 From 200 nm to the outer edge of the rise
— coastal management
— partial management

PRACTICABLE MANAGEMENT PATTERNS
1   Up to 200 nm: coastal region
2   From 200 nm to the outer edge of the rise: coastal organized area

Pattern 8
CONTINENTAL MARGIN NARROWER THAN 200 NM FROM THE BASELINES

LEGAL DATA
(i)   Territorial sea extends up to 12 nm from the baselines;
(ii) Exclusive Economic Zone has been claimed or agreed.
(iii) The coastal state cannot claim a continental shelf extended up to the outer edge of the rise, so a part of the continental margin is excluded from the national jurisdictional zone.

MANAGEMENT AREAS
1 Up to the outer edge of the rise
— coastal management
— comprehensive management

2 From the outer edge of the rise to the outer limit of the exclusive economic zone
— ocean management
— comprehensive management

The coastal state is able to manage the continental margin as a whole.

PRACTICABLE MANAGEMENT PATTERNS
1   Up to the outer edge of the rise: coastal region
2   From this edge to the outer limit of the Exclusive Economic Zone: ocean region

REGIONAL SEAS PROGRAMME: STATUS OF REGIONAL AGREEMENTS

MEDITERRANEAN ACTION PLAN
1. Convention for the Protection of the Mediterranean Sea against Pollution
Adopted: 1976
Came into force: 1978
Countries: 18, plus CEE
Approval: 3 countries
Accession: 4 countries
Ratification: 12 countries

2. Protocol for the Prevention of Pollution of the Mediterranean Sea by Dumping from Ships and Aircraft
Adopted: 1976
Came into force: 1978

3. Protocol concerning Co-operation in Combating Pollution of the Mediterranean Sea by Oil and Other Harmful Substances in Cases of Emergency
Adopted: 1976
Came into force: 1978

4. Protocol for the Protection of the Mediterranean Sea against Pollution from Land-Based Sources
Adopted: 1980
Came into force: 1983

5. Protocol concerning Mediterranean Specially Protected Areas
Adopted: 1982
Came into force: 1986

KUWAIT ACTION PLAN
1. Kuwait Regional Convention for Co-operation on the Protection of the Marine Environment from Pollution
Adopted: 1978
Came into force: 1979
Countries: 8
Ratification: 8 countries

2. Protocol concerning Regional Co-operation in Combating Pollution by Oil and Other Harmful Substances in Cases of Emergency
Adopted: 1978
Came into force: 1979

RED SEA AND GULF OF ADEN ACTION PLAN
1. Regional Convention for the Conservation of the Red Sea and Gulf of Aden Environment
Adopted: 1982
Came into force: 1985
Countries: 7 countries
Approval: 1 country
Accession: 1 country
Ratification: 5 countries

2. Protocol concerning Regional Co-operation in Combating Pollution by Oil and Other Harmful Substances in Cases of Emergency
Adopted: 1982
Came into force: 1985

WEST AND CENTRAL AFRICAN ACTION PLAN
1. Convention for Co-operation in the Protection and Development of the Marine and Coastal Environment of the West and Central African Region
Adopted: 1981
Came into force: 1984
Countries: 21
Accession: 2 countries
Ratification: 8 countries

2. Protocol concerning Co-operation in Combating Pollution in Cases of Emergency in West and Central African Region
Adopted: 1981
Came into force: 1984

EAST AFRICAN ACTION PLAN
1. Convention for the Protection, Management and Development of the Marine and Coastal Environment of the Eastern African Region
Adopted: 1985
Not yet in force
Countries: 9, plus CEE
Ratification: 1 country

2. Protocol concerning Protected Areas and Wild Fauna and Flora in the Eastern African Region
Adopted: 1985
Not yet in force

3. Protocol concerning Co-operation in Combating Marine Pollution in Cases of Emergency in the Eastern African Region
Adopted: 1985
Not yet in force

CARIBBEAN ACTION PLAN
1. Convention for the Protection and Development of the Marine Environment of the Wider Caribbean Region
Adopted: 1983
Came into force: 1986
Countries: 28 countries, plus CEE
Acceptance: 1 country
Accession: 2 countries
Ratification: 12 countries

2. Protocol concerning Co-operation in Combating Oil Spills in the Wider Caribbean Region
Adopted: 1983
Came into force: 1986

3. Protocol concerning Specially Protected Areas and Wildlife to the Convention for the Protection and Development of the Marine Environment of the Wider Caribbean Region
Adopted: 1990
Not yet in force

SOUTH-EAST PACIFIC ACTION PLAN
1. Convention for the Protection of the Marine Environment and Coastal Area of the South-East Pacific
Adopted: 1981
Came into force: 1986
Countries: 5
Ratification: 5 countries

2. Agreement on Regional Co-operation in Combating Pollution of the South-East Pacific by Hydrocarbons or Other Harmful Substances in Cases of Emergency
Adopted: 1981
Came into force: 1986

3. Supplementary Protocol to the Agreement on Regional Co-operation in Combating Pollution of the South-East Pacific by Hydrocarbons or Other Harmful Substances
Adopted: 1983
Came into force: 1987

4. Protocol for the Protection of the South-East Pacific against Pollution from Land-Based Sources
Adopted: 1983
Came into force: 1986

5. Protocol for the Conservation and Management of Protected Marine and Coastal Areas of the South-East Pacific
Adopted: 1989
Not yet in force

6. Protocol for the Protection of the South-East Pacific against Radioactive Contamination
Adopted: 1989
Not yet in force

SOUTH PACIFIC REGIONAL ENVIRONMENTAL PROGRAMME
1. Convention for the Protection of the Natural Resources and Environment of the South Pacific Region
Adopted: 1986
Not yet in force
Countries: 19
Accession: 2 countries
Ratification: 7 countries

2. Protocol concerning Co-operation in Combating Pollution Emergencies in the South Pacific Region
Adopted: 1986
Not yet in force

3. Protocol for the Prevention of Pollution of the South Pacific Region by Dumping
Adopted: 1986
Not yet in force

APPENDIX D

THE COASTAL USE FRAMEWORK (*)

category	1 SEAPORTS
sub-categories	*uses*

1.1 waterfront commercial structures	1.1.1	liquid bulk terminals
	1.1.2	solid bulk terminals
	1.1.3	multiple bulk terminals
	1.1.4	general cargo terminals
	1.1.5	lo-lo container terminals
	1.1.6	ro-ro terminals
	1.1.7	lo-lo and ro-ro terminals
	1.1.8	heavy and large cargo terminals
1.2 offshore commercial structures	1.2.1	Catenary Anchor Leg Moorings (CALMs)
	1.2.2	Exposed Location Single Buoy Moorings (ELSBMs)
	1.2.3	Single Buoy Moorings (SBMs)
	1.2.4	Single Anchor Leg Moorings (SALMs)
	1.2.5	Articulated Loading Platforms (ALPs)
	1.2.6	Mooring Towers
	1.2.7	Single Buoy Storages (SBSs)
	1.2.8	Single Anchor Leg Storages (SALSs)
	1.2.9	floating transhipment docks
	1.2.10	artificial islands
1.3 dockyards	1.3.1	ship building dockyards
	1.3.2	ship repairing dockyards
	1.3.3	oil and gas platform building dockyards
	1.3.4	oil and gas platform maintenance dockyards
1.4 passenger facilities	1.4.1	cruise ship terminals
	1.4.2	ro-ro carrier terminals
	1.4.3	hydrofoil terminals
1.5 naval facilities	1.5.1	naval dockyards
	1.5.2	surface vessel terminals
	1.5.3	submarine terminals
	1.5.4	weapon storages
	1.5.5	nuclear material deposits

236

| 1.6 | fishing facilities | 1.6.1 | vessel terminals |
| | | 1.6.2 | fishery facilities |

1.7	recreational facilities	1.7.1	sail-propelled vessel terminals
		1.7.2	engine-propelled vessel terminals
		1.7.3	wind surfing facilities
		1.7.4	yacht racing facilities
		1.7.5	semi-submersible and submarine vessel facilities

category			**2 SHIPPING, CARRIERS**
sub-categories			*uses*
2.1	bulk vessels	2.1.1	Ultra Large Crude Carriers (ULCCs)
		2.1.2	Very Large Crude Carriers (VLCCs)
		2.1.3	Medium Size Crude Carriers (MSCCs)
		2.1.4	oil product carriers
		2.1.5	chemical product carriers
		2.1.6	liquefied gas carriers (LGCs)
		2.1.7	liquefied natural gas carriers (LNGs)
		2.1.8	solid bulk carriers
2.2	general cargo vessels	2.2.1	conventional break bulk carriers
2.3	unitized cargo vessels	2.3.1	fully containerships
		2.3.2	semi-containerships
		2.3.3	ro-ro ships
		2.3.4	ro-ro cellular ships
		2.3.5	barge carriers
2.4	heavy and large cargo vessels	2.4.1	special ro-ro carriers
		2.4.2	float on-float off carriers
		2.4.3	sliding on-sliding of carriers
		3.4.4	jacking up-jacking down carriers
		3.4.5	semi-submersible carriers
		3.4.6	submersible carriers
2.5	passenger vessels	2.5.1	ro-ro carriers
		2.5.2	cruise vessels
		2.5.3	congress vessels
2.6	multipurpose vessels	2.6.1	solid bulk combined carriers
		2.6.2	solid and liquid bulk carriers
		2.6.3	bulk and general cargo carriers
		2.6.4	bulk container carriers
		2.6.5	bulk ro-ro carriers
		2.6.6	ro-ro cargo and passenger carriers

category	**3 SHIPPING, ROUTES**	
categories		*uses*
3.1 routes	3.1.1	short-sea routes
	3.1.2	feeder routes
3.2 passages	3.2.1	straits
	3.2.1	canals
3.3 separation lanes		

category	**4 SHIPPING, NAVIGATION AIDS**	
sub-categories		*uses*
4.1 buoy systems	4.1.1	simple buoys
	4.1.2	safe facilities-provided buoys
4.2 lighthouses	4.2.1	simple lighthouses
	4.2.2	radiobeacon-provided lighthouses
4.3 hyperbolic systems	4.3.1	Decca Navigator
	4.3.2	Short Range Navigation system (SHORAN)
	4.3.3	Long Range Navigation (LORAN) system
	4.3.4	Omega system
	4.3.5	Automatic Radar Plotting Aids (ARPA) system
4.4 satellite systems	4.4.1	Transit system
	4.4.2	Global Positioning system
4.5 inertial systems	4.5.1	Ship's Inertial Navigator System (SINS)

category	**5 SEA PIPELINES**	
sub-categories		*uses*
5.1 slurry pipelines	5.1.1	fine coal slurry pipelines
	5.1.2	coarse coal slurry pipelines
	5.1.3	limestone slurry pipelines
	5.1.4	phosphate slurry pipelines
	5.1.5	ore slurry pipelines
	5.1.6	copper slurry pipelines
5.2 liquid bulk pipelines	5.2.1	short submarine oil pipelines
	5.2.2	long submarine oil pipelines
	5.2.3	oil product pipelines
	5.2.4	liquefied coal pipelines

5.3 gas pipelines	5.3.1	natural gas pipelines
	5.3.2	extra natural gas pipelines
	5.3.3	coal gas pipelines
5.4 water pipelines	5.4.1	short distance aqueducts
5.5 waste disposal pipelines	5.5.1	short distance pipelines

category	**6 CABLES**

sub-categories	*uses*
6.1 electric power cables	6.1.1	low voltage cables
	6.1.2	high voltage cables

category	**7 AIR TRANSPORTATION**

sub-categories: facilities	*uses*
7.1 airports	7.1.1	waterfront-located airports
	7.1.2	offshore airports
7.2 others	7.2.1	waterfront-located heliports
	7.2.2	offshore heliports
	7.2.3	hoverports

category	**8 BIOLOGICAL RESOURCES**

sub-categories: facilities for	*uses*
8.1 fishing	8.1.1	pelagic fishing
	8.1.2	demersal fishing
	8.1.3	shellfish
8.2 gathering	8.2.1	seaweed gathering
8.3 farming	8.3.1	fish
	8.3.2	oysters
	8.3.3	mussels
	8.3.4	crustaceous
	8.3.5	seaweed
8.4 extra food products	8.4.1	marine medicine

category	9 HYDROCARBONS	
sub-categories: phases		*uses*

9.1 exploration	9.1.1	jack-up rigs
	9.1.2	semi-submersible platforms
	9.1.3	drill ships
	9.1.4	seabed wells
	9.1.5	suspended wells
9.2 exploitation	9.2.1	extended onshore drills
	9.2.2	steel framed structures (1)
	9.2.3	steel platforms (2)
	9.2.4	concrete platforms (2)
	9.2.5	hyperbuoyant leg platforms (3)
9.3 storage	9.3.1	waterfront facilities
	9.3.2	floating storages
	9.3.3	submerged storages

(1) up to 50 m depth
(2) up to the outer edge of the continental shelf
(3) beyond the outer edge of the continental shelf

category	10 METALLIFEROUS RENEWABLE RESOURCES	
sub-categories: exploitation of		*uses*

10.1 sand and gravel	10.1.1	estuarine dredging
	10.1.2	seabed dredging
	10.1.3	channel dredging
	10.1.4	maintenance dredging
10.2 water column minerals	10.2.1	salt
	10.2.2	bromine
	10.2.3	magnesium
	10.2.4	freshwater production

category	11 RENEWABLE ENERGY SOURCES	
sub-categories: environments		*uses*

11.1 wind	11.1.1	wind energy
11.2 water properties	11.2.1	thermal gradient energy
	11.2.2	salinity gradient energy
	11.2.3	density gradient energy
	11.2.4	biomass energy
	11.2.5	hydrogen production

11.3 water dynamics	11.3.1	wave power
	11.3.2	tidal power
	11.3.3	current power
11.4 subsoil	11.4.1	undersea coalmining

category	**12 DEFENCE**	
sub-categories: activity areas		*uses*
12.1 exercise areas	12.1.1	air firing exercise areas
	12.1.2	surface firing exercise areas
	12.1.3	surface bombing exercise areas
	12.1.4	submarine exercise areas
12.2 nuclear text areas		
12.3 minefields		
12.4 explosive weapon areas	12.4.1	mines
	12.4.2	torpedoes
	12.4.3	depth charges
	12.4.4	missiles

category	**13 RECREATION**	
sub-categories: areas		*uses*
13.1 onshore and waterfront	13.1.1	sunbathing
	13.1.2	swimming pools
	13.1.3	aquatic sports
	13.1.4	aquariums
13.2 offshore	13.2.1	sunbathing
	13.2.2	swimming, diving
	13.2.3	wind surfing
	13.2.4	snorkelling
	13.2.5	fishing
	13.2.6	sailing
	13.2.7	yacht racing
	13.2.8	yacht cruising
	13.2.9	housing
	13.2.10	exploration

category **14 WATERFRONT MAN-MADE STRUCTURES**

sub-categories: areas   *uses*

14.1 onshore and waterfront	14.1.1	artificial cliffs
	14.1.2	protection walls
	14.1.3	drainage basins
	14.1.4	polders
	14.1.5	artificial beaches
	14.1.6	estuarine reclamation
14.2 offshore	14.2.1	groyne fields
	14.2.2	marina breakwaters
	14.2.3	artificial reefs
	14.2.4	artificial tidal pools
	14.2.5	sea-walls
	14.2.6	jetties

category **15 WASTE DISPOSAL**

sub-categories: sources   *uses*

15.1 urban and industrial plants	15.1.1	incineration
	15.1.2	treated solid waste disposal
	15.1.3	no-treated solid waste disposal
	15.1.4	treated liquid waste disposal
	15.1.5	untreated liquid waste disposal
	15.1.6	nuclear waste disposal
15.2 watercourses	15.2.1	rivers
	15.2.2	canals
	15.2.3	underground watercourses
	15.2.4	run-off
15.3 offshore oil and gas installations	15.3.1	harmful and noxious substance discharges
	15.3.2	oil and oily mixture discharges
	15.3.3	drilling fluid discharges
	15.3.4	sewage discharges
	15.3.5	garbage discharges
15.4 dumping	15.4.1	aircraft dumping
	15.4.2	ship dumping
15.5 navigation	15.5.1	harmful and noxioussubstance discharges
	15.5.2	sewage discharges
	15.5.3	garbage discharges

category	16 RESEARCH	

sub-categories: contents		*uses*
16.1 water column	16.1.1	physics
	16.1.2	chemistry
	16.1.3	dynamics
	16.1.4	anomalous phenomena
16.2 seabed and subsoil	16.2.1	plate tectonic
	16.2.2	geology
	16.2.3	geomorphology
	16.2.4	volcanism
	16.2.5	eismicity
	16.2.6	mineralisation processes
16.3 ecosystems	16.3.1	shoreline ecosystems
	16.3.2	pelagic ecosystems
	16.3.3	demersal ecosystems
	16.3.4	food webs
16.4 external environment interaction	16.4.1	hydrological cycle
	16.4.2	sedimentary cycle
	16.4.3	erosion cycle
	16.4.4	heavy metal cycles
	16.4.5	organic substance cycles
16.5 special areas and special sensitive areas	16.5.1	estuaries
	16.5.2	coral reefs
	16.5.3	lagoons
	16.5.4	Arctic seas
	16.5.5	Antarctic seas
16.6 coastal use management	16.6.1	jurisdictional zones
	16.6.2	single use management
	16.6.3	multiple use management
	16.6.4	environmental management
	16.6.5	comprehensive management

category	17 ARCHAEOLOGY	

sub-categories: areas		*uses*
17.1 onshore and waterfront	17.1.1	harbours
	17.1.2	settlements
	17.1.3	hulls
	17.1.4	fragments of cargoes
	17.1.5	fittings

17.2 offshore	17.2.1	submerged harbours
	17.2.2	submerged settlements
	17.2.3	hulls
	17.3.4	weapons
	17.2.5	dispersed cargoes
	17.2.6	dispersed fittings

---

category   **18 ENVIRONMENTAL PROTECTION AND PRESERVATION**

---

sub-categories: areas		*uses*
18.1 onshore and waterfront	18.1.1	wetland conservation
	18.1.2	dune conservation
	18.1.3	nature reserve
	18.1.4	nature parks
	18.1.5	special areas and particularly sensitive areas
	18.1.6	ecosystem protection and preservation
	18.1.7	species conservation
18.2 offshore	18.2.1	marine reserves
	18.2.2	marine parks
	18.2.3	ecosystem protection and preservation
	18.2.4	species protection and preservation

(*) Coastal uses are numbered as in the "sea use framework", Chapter 5, Appendix 5.1.

REFERENCES: see Chapter 5, Appendix 5.1.

NOTES:
onshore: the submerged area extending landward to the coastline;
waterfront: the interface between the sea and terra firma involved by man-made structures;
offshore: the marine area extending seawards from the coastline.

APPENDIX E

THE OCEAN USE FRAMEWORK (*)

category	2 SHIPPING, CARRIERS	
sub-categories		*uses*
2.1 bulk vessels	2.1.1	Ultra Large Crude Carriers (ULCCs)
	2.1.2	Very Large Crude Carriers (VLCCs)
	2.1.3	Medium Size Crude Carriers (MSCCs)
	2.1.4	oil product carriers
	2.1.5	chemical product carriers
	2.1.6	liquefied gas carriers (LGCs)
	2.1.7	liquefied natural gas carriers (LNGs)
	2.1.8	solid bulk carriers
2.2 general cargo vessels	2.2.1	conventional break bulk carriers
2.3 unitized cargo vessels	2.3.1	fully containerships
	2.3.2	semi-containerships
	2.3.3	ro-ro ships
	2.3.4	ro-ro cellular ships
	2.3.5	barge carriers
2.4 heavy and large cargo vessels	2.4.1	special ro-ro carriers
	2.4.2	float on-float off carriers
	2.4.3	sliding on-sliding off carriers
	2.4.4	jacking up-jacking down carriers
	2.4.5	semi-submersible carriers
	2.4.6	submersible carriers
2.5 passenger vessels	2.5.1	ro-ro carriers
	2.5.2	cruise vessels
	2.5.3	congress vessels
2.6 multipurpose vessels	2.6.1	solid bulk combined carriers
	2.6.2	solid and liquid bulk carriers
	2.6.3	bulk and general cargo carriers
	2.6.4	bulk container carriers
	2.6.5	bulk ro-ro carriers
	2.6.6	ro-ro cargo and passenger carriers

category   **3 SHIPPING, ROUTES**

sub-categories		*uses*
3.1 routes	3.1.1	deep-sea routes
	3.1.2	mid-sea routes
	3.1.3	short-sea routes
	3.1.4	feeder routes
	3.1.5	mini-landbridges based routes
	3.1.6	landbridges based routes
3.2 passages	3.2.1	straits

category   **4 SHIPPING, NAVIGATION AIDS**

sub-categories		*uses*
4.3 hyperbolic systems	4.3.1	Decca Navigator
	4.3.2	Short Range Navigation system (SHORAN)
	4.3.3	Long Range Navigation (LORAN) system
	4.3.4	Omega system
	4.3.5	Automatic Radar Plotting Aids (ARPA) system
4.4 satellite systems	4.4.1	Transit system
	4.4.2	Global Positioning system
4.5 inertial systems	4.5.1	Ship's Inertial Navigator System (SINS)

category   **5 SEA PIPELINES**

sub-categories		*uses*
5.2 liquid bulk pipelines	5.2.2	long submarine oil pipelines
5.3 gas pipelines	5.3.1	natural gas pipelines

category   **6 CABLES**

sub-categories		*uses*
6.2 telephone cables	6.2.1	telephone cables
	6.2.2	facsimile system provided cables
	6.2.3	data transmission system-tied cables

category	**8 BIOLOGICAL RESOURCES**	
sub-categories: facilities for		*uses*
8.1 fishing	8.1.1 8.1.2 8.1.3	pelagic fishing demersal fishing shellfish

category	**9 HYDROCARBONS**	
sub-categories: phases		*uses*
9.1 exploration	9.1.4 9.1.5	seabed wells suspended wells

category	**10 METALLIFEROUS RESOURCES**	
sub-categories: exploitation of		*uses*
10.3 seabed deposits	10.1.1 10.1.2 10.1.3 10.1.4	phosphorite nodules manganese nodules manganese pavement nodules muds

category	**11 RENEWABLE ENERGY SOURCES**	
sub-categories: environments		*uses*
11.2 water properties	11.2.1	thermal gradient energy

category	**12 DEFENCE**	
sub-categories: activity areas		*uses*
12.1 exercise areas	12.1.1 12.1.2 12.1.3 12.1.4	air firing exercise areas surface firing exercise areas surface bombing exercise areas submarine exercise areas

12.2 nuclear test areas

12.4 explosive weapon areas	12.4.1	mines
	12.4.2	torpedoes
	12.4.3	depth charges
	12.4.4	missiles

category   **13 RECREATION**

sub-categories: areas   *uses*

13.2 offshore	13.2.5	fishing
	13.2.6	sailing
	13.2.7	yacht racing
	13.2.8	yacht cruising

category   **15 WASTE DISPOSAL**

sub-categories: sources   *uses*

15.4 dumping	15.4.1	aircraft dumping
	15.4.2	ship dumping
15.5 navigation	15.5.1	harmful and noxious substance discharges
	15.5.2	sewage discharges
	15.5.3	garbage discharges

category   **16 RESEARCH**

sub-categories: contents   *uses*

16.1 water column	16.1.1	physics
	16.1.2	chemistry
	16.1.3	dynamics
	16.1.4	anomalous phenomena
16.2 seabed and subsoil	16.2.1	plate tectonics
	16.2.2	geology
	16.2.3	geomorphology
	16.2.4	volcanism
	16.2.5	seismicity
	16.2.6	mineralisation processes
16.3 ecosystems	16.3.2	pelagic ecosystems
	16.3.3	demersal ecosystems
	16.3.4	food webs

16.4 external environment interaction	16.4.1	hydrological cycle
	16.4.4	heavy metal cycles
	16.4.5	organic substance cycles
16.5 special areas and particularly sensitive areas	16.5.4	Arctic seas
	16.5.5	Antarctic seas
16.6 ocean management	16.6.1	jurisdictional zones
	16.6.2	single use management
	16.6.3	multiple use management
	16.6.4	environmental management
	16.6.5	comprehensive management

category	**17 ARCHAEOLOGY**	
sub-categories: areas		*uses*
17.2 offshore	17.2.5	dispersed cargoes

category	**18 ENVIRONMENT PROTECTION AND PRESERVATION**	
sub-categories: areas		*uses*
18.2 offshore	18.2.3	ecosystem protection and preservation
	18.2.4	species protection and preservation

(*) Ocean uses are numbered as in the "sea use framework" (Chapter 4, Appendix 4.2)

REFERENCES: see Chapter 4, Appendix 4.2.

INDEX

251

Milton Keynes UK
Ingram Content Group UK Ltd.
UKHW040109071024
449327UK00019B/922